C0-ANN-492

EOCENE MOLLUSCAN BIOSTRATIGRAPHY OF THE PINE MOUNTAIN AREA, VENTURA COUNTY, CALIFORNIA

BY
CHARLES R. GIVENS

UNIVERSITY OF CALIFORNIA PUBLICATIONS IN GEOLOGICAL SCIENCES
Volume 109

UNIVERSITY OF CALIFORNIA PRESS

EOCENE MOLLUSCAN BIOSTRATIGRAPHY OF THE PINE MOUNTAIN AREA, VENTURA COUNTY, CALIFORNIA

EOCENE MOLLUSCAN BIOSTRATIGRAPHY OF THE PINE MOUNTAIN AREA, VENTURA COUNTY, CALIFORNIA

BY

CHARLES R. GIVENS

UNIVERSITY OF CALIFORNIA PRESS
BERKELEY · LOS ANGELES · LONDON
1974

UNIVERSITY OF CALIFORNIA PUBLICATIONS IN GEOLOGICAL SCIENCES

Volume 109

Approved for publication June 22, 1973
Issued April 24, 1974

UNIVERSITY OF CALIFORNIA PRESS
BERKELEY AND LOS ANGELES
CALIFORNIA

UNIVERSITY OF CALIFORNIA PRESS, LTD.
LONDON, ENGLAND

ISBN: 0-520-09506-5
LIBRARY OF CONGRESS CATALOG CARD NUMBER: 73-620160

PRINTED IN THE UNITED STATES OF AMERICA

CONTENTS

EOCENE MOLLUSCAN BIOSTRATIGRAPHY OF THE PINE MOUNTAIN AREA, VENTURA COUNTY, CALIFORNIA

BY

CHARLES R. GIVENS

ABSTRACT

Approximately 14,000 ft of marine, deltaic, and continental Eocene strata are exposed in the Pine Mountain area, northern Ventura County, California. These strata are assigned to four partly intertonguing formations: Juncal Formation, Matilija Sandstone, Cozy Dell Shale, and Coldwater Sandstone. The Juncal Formation, approximately 12,000 ft thick, forms the major portion of the Eocene section. It is composed of interbedded and complexly intertongued arkosic sandstone, cobble conglomerate, sandy siltstone, and mudstone. It lies unconformably on granitic and metamorphic rocks of the pre-Tertiary Basement Complex along the eastern margin of the map area. The Juncal Formation is interpreted to be largely of deltaic and marine origin, although much of the conglomerate was probably deposited in a continental environment. The Matilija Sandstone overlies and, in part, intertongues with the Juncal Formation. It is composed predominantly of fine-grained, well-sorted, marine sandstone and attains a maximum thickness of about 1,600 ft within the map area. The Cozy Dell Shale occurs as an eastward-lensing tongue within the Matilija Sandstone. It is composed of marine mudstone with thin sandstone interbeds and has a maximum thickness of about 700 ft. The Coldwater Sandstone conformably overlies the Matilija Sandstone. Only a partial section of the formation, about 700 ft thick, has been preserved. The unit is composed mainly of thick-bedded, well-sorted, medium- to coarse-grained, arkosic sandstone. Lithologic features and fossils indicate that it was probably deposited in very shallow marine and brackish-water environments.

Fossil mollusks are abundant in certain parts of the Pine Mountain section. Collections have been obtained from 78 localities and a total of 194 molluscan taxa, including 78 pelecypods, 113 gastropods, and 3 scaphopods, have been identified. Ten of these taxa represent new or questionably new species.

Four stratigraphically distinct molluscan faunas are recognized in the Pine Mountain section. From oldest to youngest, they are: *Turritella uvasana infera* fauna, *Turritella uvasana applinae* fauna, *Ectinochilus supraplicatus* fauna, and *Ectinochilus canalifer* fauna. The first three faunas occur within the Juncal Formation. The *Ectinochilus canalifer* fauna includes the mollusks in the Matilija Sandstone, Cozy Dell Shale, Coldwater Sandstone, and the upper 1,000 ft of the Juncal Formation. These faunas are closely related and overlap in taxonomic composition; each, however, contains many taxa that (in the Pine Mountain section) are restricted to it.

The molluscan faunas in the Pine Mountain section are similar in composition to those that characterize the Eocene "Capay," "Domengine," "Transition," and "Tejon" "Stages" of Clark and Vokes (1936). The faunal sequence, furthermore, confirms the chronological sequence of these "Stages." Each "Stage" can be confidently recognized by the restricted occurrence of some taxa and the concurrent ranges of others. Several taxa formerly regarded as "index fossils" for certain "Stages," however, range into younger or older "Stages" in the Pine Mountain section and, therefore, are of little value for precise age determination. In particular, the association in this section of the "Upper Capay" index fossil *Galeodea susanae* with taxa characteristic of the "Domengine Stage" indicates that the upper part of the "Capay Stage" (i.e., the *Galeodea susanae* Zone of Clark and Vokes, 1936) and the "Domengine Stage" are equivalent.

Although the faunal data from the Pine Mountain section confirm the existence and sequence of Clark and Vokes's Eocene "Stages," these units are in need of further revision and refinement before they can be accepted as formal chronostratigraphic units. Their nomenclature is unsatisfactory. Most of them have been named after formations. In order to avoid confusion between lithostratigraphic units and chronostratigraphic units, new names should be proposed for these "Stages." A more serious shortcoming of the "Stages" is their lack of precise, chronologically significant, boundaries. Previous authors have generally equated the boundaries of these units with the boundaries of the formations in which they are typically developed, even though the characteristic fauna of a given "Stage" may be confined to only a portion of the corresponding formation. The various assemblage-zones assigned to these "Stages" by previous authors are also unsatisfactory for defining Stage boundaries because their limits are poorly defined. It is suggested that future biostratigraphic studies of Eocene mollusks on the Pacific Coast be directed toward establishing a zonation based upon evolutionary events and that the limits of these Zones be used to define Stage boundaries. Apparent evolutionary sequences are recognized within several molluscan taxa in the Pine Mountain section.

INTRODUCTION

FOSSIL MOLLUSKS are abundant and widely distributed in marine Eocene strata on the Pacific Coast of North America and have received considerable study during the last hundred years. The faunas of many formations have been described and the general faunal sequence within the Eocene has been reasonably well worked out. The details of this sequence and the systematic and phylogenetic relationships of a majority of the molluscan taxa, however, are poorly understood. A major impediment to more refined stratigraphic and paleontologic studies of Eocene mollusks on the Pacific Coast is the lack of a well-defined and well-substantiated system of chronostratigraphic units for correlating fossiliferous sections.

Previous attempts at correlation of megafossiliferous Eocene strata have relied mainly upon the "Stages"[1] (fig. 6) of Clark and Vokes (1936). These "Stages," based upon the molluscan faunas of certain formations that presumably characterize different parts of the Eocene, evolved in a rather haphazard manner. Adequate superpositional control is lacking for most of them and their boundaries are poorly defined. The various "Zones" assigned to these "Stages" by Clark and Vokes (1936) and by Weaver et al. (1944) represent little more than local assemblage-zones characterized by certain "index fossils." Few of them have been precisely defined. In particular, the molluscan fauna of the "Transition Stage" or "*Rimella supraplicata* Zone" is largely undescribed, making this presumed time-rock interval exceedingly difficult to recognize. Because of the uncertain stratigraphic relationships of these "Stages" and "Zones," correlations based on them are tenuous and imprecise at best.

Mallory (1959) has established a sequence of well-defined Stages and Zones for the Lower Tertiary of the Pacific Coast based upon benthonic foraminiferal assemblages. Although these units are suitable for correlation of fine-grained offshore marine deposits in which foraminifera are often abundant, they are difficult to recognize in nearshore mollusk-bearing strata because of the scarcity of foraminifera in these deposits.

Because of the need for biostratigraphic data that would permit more precise and reliable correlation of megafossiliferous Eocene strata on the Pacific Coast than is presently possible, this study was undertaken. Its primary objective is to provide a detailed analysis of the Eocene molluscan faunal sequence within a local area. The Pine Mountain area was chosen because the faunal sequence

[1] The "Stages" of Clark and Vokes (1936) are placed in quotation marks throughout the text to indicate that their usage is informal and to distinguish them from Formations of the same name.

there is more complete than at any other known Eocene section on the Pacific Coast. Molluscan assemblages referable to the "Meganos," "Capay," "Domengine," "Transition," and "Tejon" "Stages" have been reported (Schlee, 1952; Hartman, 1957; Kiessling, 1958; Newman, 1959; Welday, 1960; Jestes, 1963) from various parts of the area, but no detailed systematic or stratigraphic study of these assemblages has been made. Other features of the Pine Mountain section further enhance its suitability for this study: the Eocene strata are well exposed; the section is conformable and uninterrupted by major structural complications; and fossil mollusks occur at various stratigraphic horizons from near the base to near the top of the section.

The fieldwork was carried out over a period of six months during the summer of 1966 and the spring and fall of 1967. Emphasis in the field was focused primarily on working out the stratigraphic relationships between the various rock units and on collecting fossils from as many different stratigraphic horizons in the section as possible. The fossil collections were subsequently studied in the Department of Geological Sciences of the University of California at Riverside. The geologic map (in pocket) represents a compilation of my own mapping plus parts of several unpublished theses maps that were prepared mainly by graduate students of the University of California, Los Angeles, under the supervision of Professor John C. Crowell. The portions of the area mapped by me and by the authors of the theses are indicated on the geologic map. The map covers parts of seven quadrangles: Lockwood Valley, Topatopa Mountains, Devils Heart Peak, San Guillermo, Lion Canyon, Reyes Peak, and Wheeler Springs.

The results of this study confirm the existence and sequence of the traditional Eocene "Stages" ("Capay," "Domengine," "Transition," and "Tejon") of the Pacific Coast. Each "Stage" can be confidently recognized by the restricted occurrence of some molluscan taxa and the concurrent ranges of others. A few taxa formerly regarded as "index fossils" for certain "Stages," however, range into older or younger "Stages" in the Pine Mountain section. In particular, the association in this section of the "Upper Capay" index fossil *Galeodea susanae* with taxa characteristic of the "Domengine Stage" indicates that the upper part of the "Capay Stage" (i.e., the *Galeodea susanae* Zone of Clark and Vokes, 1936) and the "Domengine Stage" are equivalent. Thus, only the lower of the two faunal zones assigned to the "Capay Stage" by Clark and Vokes (1936) and others (Vokes, 1939; Weaver, et al., 1944) is referable to this "Stage." Further revision of the Eocene "Stages" is necessary before they can be accepted as formal chronostratigraphic units. Their nomenclature is unsatisfactory. Most of them have been named after formations, a practice that is undesirable because it tends to obscure the distinction between lithostratigraphic and chronostratigraphic units. Therefore, new geographic names should eventually be proposed for the "Stages." The boundaries of these units are also poorly defined. Previous authors have generally equated their boundaries with formation boundaries, even though the characteristic fauna of a given "Stage" may be confined to only a small portion of the corresponding formation. Precise and chronologically significant biostratigraphic limits have not been established for these "Stages." It is suggested that future work be directed toward the establishment of a zonation for the Pacific Coast Eocene based on evolutionary events within certain molluscan lineages and that the limits of these zones be used to define formal Stage boundaries. Apparent evolutionary sequences have been recognized within several molluscan taxa in the Pine Mountain section.

Acknowledgments

I am indebted to Dr. Michael A. Murphy of the Department of Geological Sciences, University of California, Riverside, for advice and encouragement during all phases of this investigation and for critically reading the manuscript.

I am especially grateful to Dr. Murphy and to William J. Zinsmeister for photographing most of the fossils. Mrs. Carole S. Hickman, of Swarthmore College, also provided use of her photographic facilities and assisted in the photographing of some of the fossils.

I would also like to thank Edward E. Welday, Edwin C. Jestes, John S. Schlee, and Edmund W. Kiessling for permission to use data from their unpublished theses maps.

Financial support for the early phases of this investigation was provided by two National Science Foundation Traineeships and a University of California Fellowship. The Academy of Natural Sciences of Philadelphia provided financial support during the final phase of the project. Field expenses were partly alleviated by grants from the Hewett Fund of the Department of Geological Sciences, University of California, Riverside.

PREVIOUS INVESTIGATIONS IN THE PINE MOUNTAIN AREA

There is little published information on the geology of the Pine Mountain area. The general distribution of the Eocene strata in this region is shown on the 1:250,000 Los Angeles Sheet of the Geologic Map of California (Jennings and Strand, 1969). A more detailed but still somewhat generalized map (scale, 1:125,000) was published by Merrill (1954). Merrill assigned the Eocene strata in the Pine Mountain area to the Juncal, Matilija, and Cozy Dell formations. Dickinson (1969:15–17) discussed the stratigraphy of the Eocene deposits in the western half of the Pine Mountain area. He assigned the strata to the Matilija, Cozy Dell, and Coldwater formations, pointed out the intertonguing relationship between the Matilija and Cozy Dell formations, and subdivided the Matilija Sandstone into two lithologic phases: the "Reyes Phase," composed of massive and cross-bedded, gray to white, medium- to coarse-grained, locally conglomeratic sandstone; and the "Derrydale Phase," composed of laminated and cross-laminated, green to tan, fine- to medium-grained sandstone.

Most of the information on the geology of the Pine Mountain area is contained in unpublished graduate student theses. Gazin (1930) prepared a small-scale geologic map of a large area north and east of Pine Mountain and was the first to note the presence of Eocene fossils in this area. Dreyer (1935) mapped the upper Sespe Creek area, including a portion of Pine Mountain. From strata exposed on the south slope of Pine Mountain, he collected a small molluscan assemblage which he considered to be "upper Domengine" in age and which Clark and Vokes (1936) subsequently referred to the "Transition Stage."

Following Dreyer's and Gazin's studies, which were essentially reconnaissance in nature, more detailed studies of the Pine Mountain area were carried out by other graduate students, including Schlee (1952), Hartman (1957), Kiessling (1958), Newman (1959), Poyner (1960), Welday (1960), and Jestes (1963). These geologists prepared large-scale geologic maps that illustrate in considerable detail the complex stratigraphic relationships of the Eocene units in this area. These detailed maps made the present study possible. Schlee and Kiessling mapped parts of the Lockwood Valley and Topatopa Mountains quadrangles in the eastern third of the Pine Mountain area, where the Eocene strata are in contact with the crystalline rocks of the Basement Complex. Near Sespe Hot Springs, Schlee collected fossil mollusks which he assigned to the "Meganos" and "Capay" "Stages." Kiessling reported mollusks characteristic of the "Transition Stage" from the Eocene strata near Thorn Meadows and Grade Valley. Welday mapped a portion of the San Guillermo quadrangle, northwest of Kiessling's area, and identified mollusks that he considered to be characteristic of the "Domengine" and "Transition" "Stages." Hartman, Poyner, and Newman mapped areas adjacent to the Big Pine Fault, along the northern margin of the Pine Mountain area. Newman identified mollusks that he considered to be characteristic of the "Transition Stage" from Eocene strata near the confluence of Alamo Creek and Dry Canyon. Jestes, as part of a regional stratigraphic study of Eocene sandstones in the western Transverse Ranges, mapped a large part of the Pine Mountain area and synthesized the earlier geologic studies. In the western part of the area, he collected fossil mollusks characteristic of the "Tejon Stage."

GEOLOGIC SETTING AND GENERAL STRUCTURAL FEATURES

The Pine Mountain area is located in northern Ventura County, near the northern boundary of the Transverse Ranges and about 12 mi southwest of the intersection of the Big Pine and San Andreas faults (fig. 1). The Eocene strata crop out in a roughly triangular area bounded on the north and south by faults and on the east by igneous and metamorphic rocks of the pre-Tertiary Basement Complex (Bailey and Jahns, 1954:86). They lie unconformably on the basement rocks and are overlain unconformably by Quaternary terrace gravels and by a fanglomerate deposit of probable Pleistocene age (Welday, 1960:79).

The major structures of the Pine Mountain area include the Big Pine, Pine Mountain, and San Guillermo faults and several large folds. The Big Pine fault (Hill and Dibblee, 1953), bounding the area on the northwest, is a major left-lateral strike-slip fault of the Transverse Ranges. Hill and Dibblee (1953:452) and Poyner (1960) have presented evidence for 8–10 mi of left-lateral movement along this fault. This movement has brought Miocene and younger continental deposits on the north side of the fault into juxtaposition with the Eocene strata on the south side (James, 1963).

The Pine Mountain fault, bounding the area on the south, is a north-dipping reverse fault (Hill, 1954:10) along which the Eocene strata of the Pine Mountain area have been elevated thousands of feet relative to Eocene and younger strata in the Sespe Creek region south of the fault. The Pine Mountain fault has a northwest trend similar to that of the Nacimiento fault in the southern

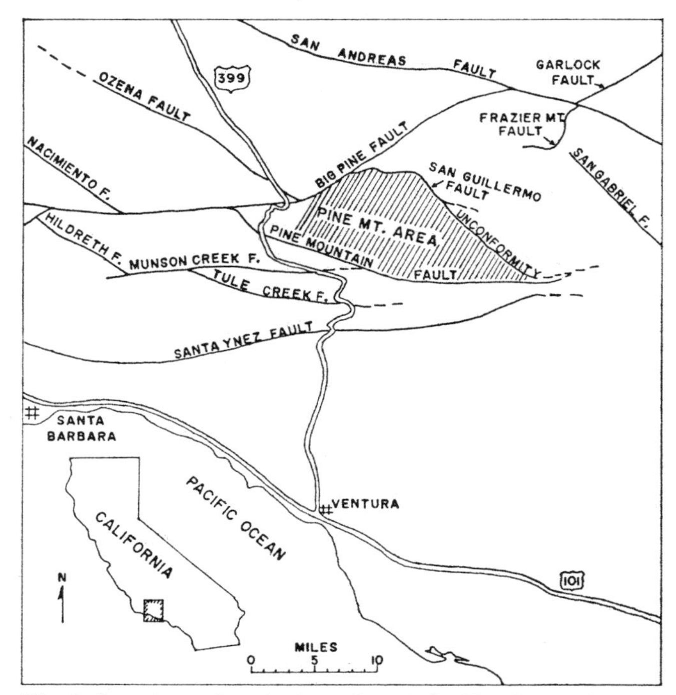

Fig. 1. Location and geologic setting of the Pine Mountain area.

Coast Ranges, and Vedder and Brown (1968:254–256) have suggested that it may represent a southeastward continuation of this fault which has been displaced about 10 mi to the east by movement along the Big Pine fault.

The San Guillermo fault (Poyner, 1960; Welday, 1960; Carman, 1964) bounds the area on the north and northeast. It is a west-dipping, generally northwest-trending fault which exhibits an apparent reversal of throw along its trace. North and west of San Guillermo Mountain, it appears to have a reverse sense of displacement with Eocene strata on the hanging wall abutting against Miocene? and younger continental deposits on the footwall (Poyner, 1960:60–62; Welday, 1960:87, 88). To the southeast in the vicinity of Grade Valley, however, the Eocene rocks on the hanging wall are in contact with older granitic rocks on the footwall and the fault appears to have a normal sense of displacement. Such apparent reversals in the sense of displacement are not uncommon along faults that have undergone principally strike-slip movement, and Welday (1960:90–93) has presented evidence in favor of significant right-lateral strike-slip movement along the San Guillermo fault. According to Welday, the Pleistocene? fanglomerate that underlies San Guillermo Mountain appears to have been offset several miles in a right-lateral sense from its probable source area in the Basement Complex to the southeast. The San Guillermo fault is truncated on the north by the Big Pine fault. Poyner (1960:64) suggested that the abrupt westward bend in the trace of the San Guillermo fault north of San Guillermo Mountain may owe to drag along the Big Pine fault. Poyner (1960; 1965:1088) also presented evidence that the Ozena fault in the Sierra Madre Range, north of the Big Pine fault, is probably the offset northwestward continuation of the San Guillermo fault. The southeastward extent of the San Guillermo fault is unknown. East of Grade Valley, in the vicinity of the axis of the Grade Valley anticline, it continues beyond the mapped area into the Basement Complex. South of the Grade Valley anticline, the Eocene strata lie unconformably on the Basement Complex.

Within the mapped area, the Eocene strata have been deformed into a series of west- to northwest-trending, slightly asymmetrical folds. Four of these folds—the Pine Mountain anticline, Piedra Blanca syncline, Grade Valley anticline, and San Guillermo syncline—are large structures that extend across the area and are truncated by the bounding faults. Smaller folds occur adjacent to the Pine Mountain fault, south and east of Wegis Ranch, and in the vicinity of Thorn Meadows. Most of the folds plunge gently toward the west or northwest, except near the Pine Mountain fault where they have a southeastward plunge. In the eastern part of the area, near Thorn Meadows and Thorn Point, the entire Eocene section is exposed in a southwestward dipping homocline.

The Eocene strata are locally offset by faults of relatively small magnitude. Displacements along most of these faults range from a few tens of feet to a few hundred feet. Poyner (1960:73–77), however, presented evidence for left-lateral offsets of more than 1,000 ft along the three northeast-trending faults northwest of San Guillermo Mountain. He considered these faults to have formed in response to movement along the Big Pine fault. The two west-trending faults south of Wegis Ranch may also be related in origin to the Big Pine fault. In the central portion of the mapped area, the Eocene strata are offset by a northeast-trending fault that appears to have formed in response to folding of the strata. This fault exhibits its greatest displacement midway between the axes of the Piedra Blanca syncline and Grade Valley anticline and dies out in the vicinity of the fold axes. Because of pronounced facies changes in the Eocene strata, the total displacement along this fault is difficult to determine but is estimated to be at least several hundred feet.

STRATIGRAPHY

Approximately 14,000 ft of Eocene strata are exposed in the Pine Mountain area. The section is entirely conformable and has been subdivided into four formations: Juncal Formation, Matilija Sandstone, Cozy Dell Shale, and Coldwater Sandstone. The lithologic characteristics and stratigraphic relationships of these formations are summarized in figure 2. The Eocene strata lie unconformably on crystalline rocks of the pre-Tertiary Basement Complex along the eastern margin of the area and are unconformably overlain by Quaternary terrace deposits and by a fanglomerate unit of probable Pleistocene age. Deposits of Oligocene, Miocene, or Pliocene age have not been recognized within the mapped area.

JUNCAL FORMATION

A thick unit of interbedded and intricately intertongued sandstone, conglomerate, siltstone, and mudstone which forms the major portion of the Eocene section in the Pine Mountain area is here referred to the Juncal Formation of

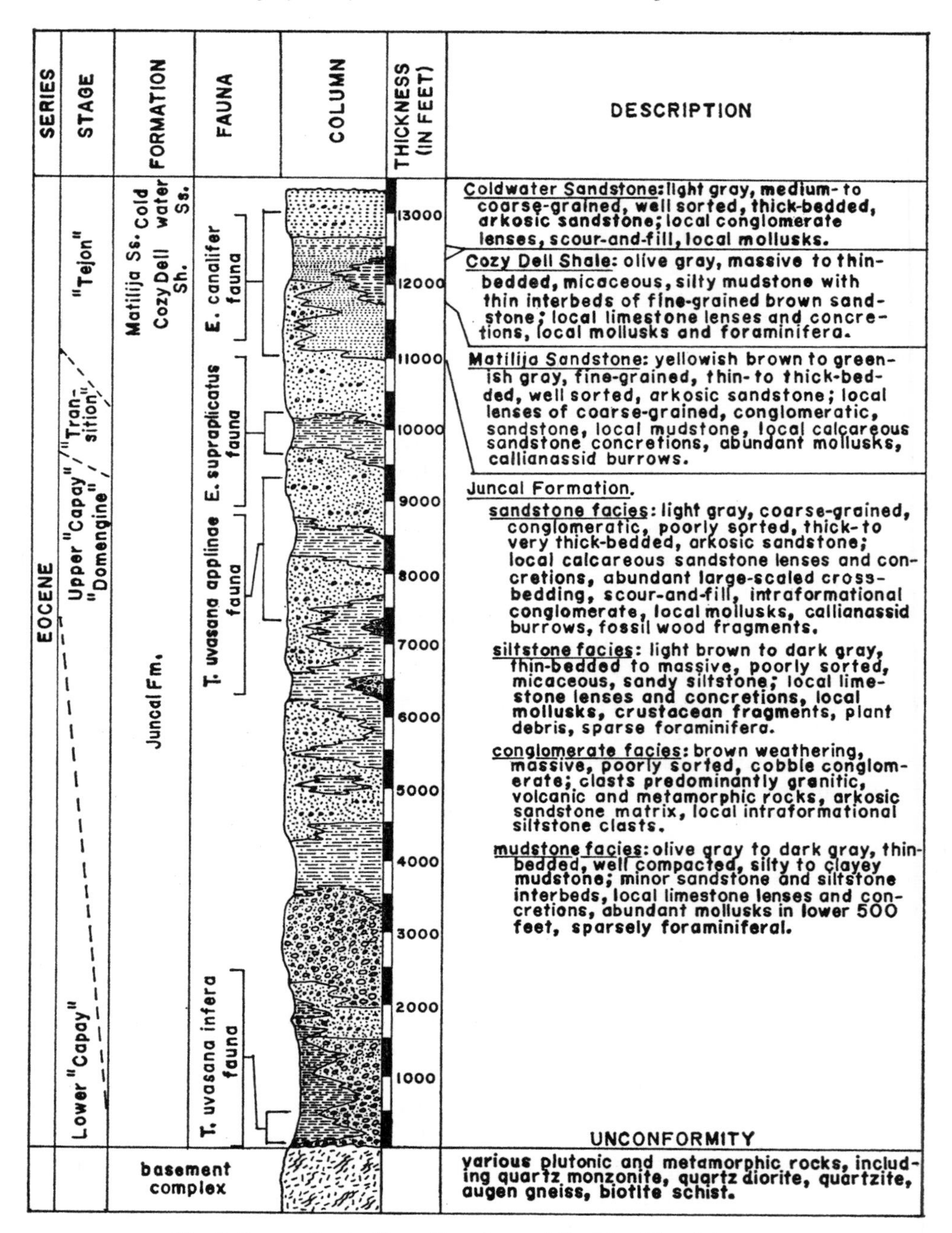

Fig. 2. Composite stratigraphic column of the Pine Mountain area.

Page et al. (1951:1749). Except for the local occurrence of thick conglomerates, this unit is lithologically similar to the Juncal Formation elsewhere in the Transverse Ranges and occupies a similar stratigraphic position within the Eocene sequence. Most of these strata have also been assigned to the Juncal Formation by Merrill (1954) and Dickinson (1969:13). In the western half of the mapped area, however, light gray sandstone here included in the upper part of the Juncal Formation has been referred by Merrill (1954) and Dickinson (1969:15) to the Matilija Sandstone. Various informal names have also been applied to the strata that I refer to the Juncal Formation, including Mutau Formation (Schlee, 1952), Thorn Meadows Beds (Kiessling, 1958), Thorn Meadows Formation (Jestes, 1963), and Park Canyon Formation (Welday, 1960).

The type section of the Juncal Formation is located near Jameson Lake in the upper Santa Ynez River Valley, about 15 mi southwest of Reyes Peak in the Pine Mountain area. At the type section, the formation is about 3,500 ft thick and is composed of interbedded sandstone and shale, with shale predominating (Page et al., 1951:1749; Dibblee, 1966:22). It lies unconformably on Cretaceous strata and is conformably overlain by the Matilija Sandstone. Locally in the type area, the Juncal Formation is conformably underlain by the Sierra Blanca Limestone (Page et al., 1951:1753; Dibblee, 1966:20).

The Juncal Formation in the Pine Mountain area is approximately 12,000 ft thick. It lies unconformably on granitic and metamorphic rocks of the pre-Tertiary Basement Complex along the eastern margin of the area. The unconformity is well exposed in the vicinity of Sespe Hot Springs (geologic map) where a conglomerate bed, about 5 ft thick, is in depositional contact with the basement rocks. The conglomerate is composed mainly of subrounded cobbles and boulders of quartz monzonite, similar to that which is exposed beneath the unconformity. The Juncal Formation intertongues at the top with the Matilija Sandstone. The contact is gradational and has been drawn at the transition from typical light gray, medium- to coarse-grained, poorly sorted, conglomeratic sandstone of the Juncal Formation to greenish gray and yellowish brown, fine- to medium-grained, well-sorted sandstone of the Matilija Sandstone. The contact is well exposed in the cliffs southeast of Wegis Ranch, in the western half of the mapped area.

The Juncal Formation is composed of four principal rock types: light gray arkosic sandstone; cobble- and pebble-conglomerate; gray to brown sandy siltstone; and gray mudstone. These four rock types have been mapped as separate units appropriate to the scale of the map and are treated herein as informal lithofacies of the formation. The sandstone and siltstone facies make up the major portion of the formation. The conglomerate and mudstone facies are confined to the lower half of the formation in the eastern part of the mapped area.

Sandstone facies.—This facies is composed predominantly of light gray to yellowish gray, locally conglomeratic, arkosic sandstone, which occurs throughout the Juncal Formation as numerous tongues and lenses interbedded with the other rock types. It is well exposed in the vicinity of Thorn Meadows in the eastern half of the mapped area. The sandstone ranges from medium- to very coarse-grained, with coarse-grained beds predominant, and is typically poorly sorted and thick- to very thick-bedded. The bedding is markedly lenticular and abrupt lensing of the sandstone into the other lithofacies, particularly the siltstone, is common. Petrographic examination of the sandstone indicates that it is composed mainly of subangular to subrounded grains of quartz and feldspar, with small amounts of mica, lithic fragments, and minor accessory minerals. The matrix consists mainly of fine-grained sand and silt; little argillaceous material is present. Locally, the sandstone is well cemented with calcium carbonate and/or iron oxide cement. Characteristic sedimentary structures of the sandstone include large-scale cross-stratification, scour-and-fill and intraformational conglomerate. The cross strata are of both *trough* and *planar* types and occur in sets ranging from 3 to 15 ft in thickness. The intraformational conglomerate consists mainly of angular to subrounded clasts of gray to brown siltstone similar to that of the siltstone facies of the Juncal Formation. The clasts range from 1 in to 3 ft in thickness but most are less than 12 in thick. Locally, fossil wood fragments up to a foot in length occur in the sandstone.

Numerous conglomerate lenses occur throughout the sandstone facies. These lenses range from 3 to 100 ft in thickness and are similar in appearance to the conglomerate facies of the Juncal Formation. They are composed predominantly of well-rounded cobbles and pebbles of igneous (granitic and volcanic) and metamorphic (chiefly quartzite) rocks.

Hard, dark brown–weathering, calcareous sandstone concretions also occur throughout the sandstone facies and are abundant at certain horizons. These concretions are spheroidal to lenticular in shape and range from 1 to 8 ft in thickness, although most are less than 3 ft thick.

Locally, the sandstone is honeycombed with distinctive branching burrows that are strikingly similar to those made by the decapod crustacean *Callianassa major* Say, the common "ghost shrimp" of the intertidal zone along the southeastern Atlantic Coast of the United States. The burrowing activities of this species have been described by Weimer and Hoyt (1964). The burrows in the sandstone facies range from 1 in to 1.5 in in diameter and up to 3 ft in length. They occur in various orientations in the sandstone but the longest burrows are nearly perpendicular to bedding. Characteristically, the burrows are lined with a thin layer of clay, about ⅛ in thick, which has a nodular outer surface and a smooth inner surface. Similar features characterize the burrows of *Callianassa major* (Weimer and Hoyt, 1964; Hoyt and Weimer, 1965). The inference that these burrows were made by a callianassid crustacean similar to *C. major* is supported by the occurence of callianassid claws at several localities (e.g., UCR localities 4700 and 4703) in the siltstone facies of the Juncal Formation.

In the upper 2,000 ft of the sandstone facies, thin to thick beds of greenish gray to buff, moderately sorted, fine- to medium-grained sandstone are locally interbedded with the light gray, coarse-grained, conglomeratic sandstone typical of this facies. This fine- to medium-grained sandstone is most abundant in the southern and western portions of the mapped area. Toward the north and east, it grades into coarse-grained, conglomeratic sandstone. In the San Guillermo syncline, east of Alamo Creek, the upper part of the sandstone facies is composed almost entirely of light gray, coarse-grained, locally conglomeratic sandstone.

Associated with the fine- to medium-grained sandstone beds in the upper part of the sandstone facies are a few beds of red and green mudstone. Some of these beds, which range from 2 to 6 ft in thickness, can be traced for considerable distance along strike. One such interval of red and green mudstone, 10 to 15 ft thick, has been mapped as a marker horizon for structural control in the vicinity of Alamo Creek and the San Guillermo syncline (see geologic map). Some of the green mudstone beds contain abundant oysters.

The upper part of the sandstone facies in the western part of the mapped area has been assigned by Dickinson (1969:15) to the Matilija Sandstone of Kerr and Schenck (1928:1090). I have referred these beds to the Juncal Formation because they are lithologically similar to, and physically continuous with, the sandstone tongues that occur throughout this formation in the eastern part of the mapped area.

Conglomerate facies.—This facies is composed of several lenticular and tongue-haped conglomerate units that occur in the lower half of the Juncal Formation ıorth and east of Thorn Meadows. The conglomerate is interbedded with the sand-tone and siltstone facies and intertongues toward the southeast with the mud-one facies. Northeast of Thorn Meadows, the lowest conglomerate occurs at the ıse of the Juncal Formation and lies unconformably on granitic rocks of the ısement Complex. The conglomerate units vary considerably in thickness. The o largest units, east of Thorn Meadows, are reported to have maximum thick-sses of about 1,100 ft (lower unit) and 1,800 ft (upper unit), respectively chlee, 1952; Kiessling, 1958). The small lenses near Grade Valley, on the other nd, range from about 200 to 400 ft in thickness. Most of the conglomerate its are less than 1,000 ft thick.

The conglomerate is composed predominantly of well-rounded clasts of igneous d metamorphic rocks. It is massive to indistinctly very thick-bedded and is ıerally poorly sorted. In most of the outcrops examined, cobbles, pebbles and all boulders are scattered through an abundant arkosic sandstone matrix. ally, however, the conglomerate is well sorted and composed predominantly closely packed cobble- or pebble-sized clasts. In some outcrops, imbrication of clasts was noted. The sandstone matrix of the conglomerate is similar in lith-gy and texture to that of the sandstone facies of the formation.

'he following rock types have been identified in the conglomerate: granite, rtz monzonite, quartz diorite, pegmatite, aplite, gabbro, andesite, basalt, rhyo-?, biotite schist, augen gneiss, and sandstone. The granitic rocks and the rtzite are the most abundant clasts, but volcanic rocks are also common. ally, angular to subrounded clasts of gray siltstone up to 3 ft in diameter present in the conglomerate. These clasts are apparently of intraformational in. They are lithologically similar to the siltstone facies of the Juncal Forma- and are most abundant where this facies is directly overlain by the con-rate.

stone facies.—This facies is composed predominantly of dark gray to light sandy siltstone with subordinate sandstone interbeds. In the north-central of the mapped area, it forms the major part of the Juncal Formation and has a minimum thickness of about 5,000 ft. The base of the siltstone is faulted and its total thickness may be much greater. Toward the south and east, it intertongues with the other lithofacies of the formation.

The siltstone varies from dark gray to light brown on fresh surface and weathers yellowish gray to yellowish brown. It is typically sandy, micaceous, massive to indistinctly thin-bedded, and poorly sorted. Fissility is generally poorly developed, although occasional thinly laminated, moderately fissile layers are present. The sand content of the siltstone varies considerably and much of this facies may be more appropriately classified as very fine-grained silty sandstone. Jarosite and gypsum are abundant on weathered surfaces of the siltstone and probably account for the slight sulfureous odor given off when the rock is freshly broken. Locally, fossil plant remains are abundant. Pieces of wood from a few inches to as much as a foot in length are particularly common. Hartman (1957:14) reported the discovery of "two fossil palm trunks and much fossil palm debris" in association with black carbonaceous shale layers in the siltstone near the San Guillermo Fault and also noted (ibid., p. 15) that "many of the surrounding sandstone layers contain abundant limonitized plant fragments; these are unidentified bits of stems, stalks and leaves impressed in the bedding planes."

Numerous lenses and concretions of very fine-grained, silty limestone, 3 in to 3 ft thick, occur throughout the siltstone facies and are abundant at certain horizons. The limestone is dark gray to black on fresh surface and weathers light gray to pale orange. Locally, these limestone lenses and concretions contain well-preserved fossil mollusks.

Sandstone interbeds within the siltstone are mainly of two types: yellowish gray to white, thick-bedded, coarse-grained, locally conglomeratic sandstone similar to that of the sandstone facies of the formation; and gray to brown, thin- to thick-bedded, fine- to medium-grained, moderately well-sorted sandstone. The light gray conglomeratic sandstone is most common in the vicinity of tongues of the sandstone facies. The thin-bedded, fine- to medium-grained sandstone beds occur throughout the siltstone facies but are particularly abundant in the portion of this facies exposed between Wagon Road Creek and the Big Pine fault, along the northern margin of the mapped area.

Mudstone facies.—This facies is composed predominantly of silty to clayey mudstone with minor interbeds of sandy siltstone and sandstone. It is confined to the lower half of the Juncal Formation in the extreme eastern portion of the mapped area, where it intertongues with the sandstone and conglomerate facies. Northwest of Mutau Flat, it lenses entirely into these facies. Along the eastern boundary of the map, between Sespe Hot Springs and Piru Creek, the mudstone facies lies unconformably on crystalline rocks of the Basement Complex. The actual contact with the basement rocks is poorly exposed but is interpreted to be unconformable because it parallels the strike of the mudstone facies and because the mudstone intertongues laterally with sandstone and conglomerate units that lie unconformably on the basement rocks. The total thickness of the mudstone facies is uncertain. It is estimated to be at least 4,000 ft thick at the southern margin of Mutau Flat. It may be considerably thicker in the area south of Johnston Ridge, but the combination of folding, facies changes, and dense brush cover in that area make thickness estimates unreliable.

The mudstone varies from olive gray to dark gray on fresh surface and weathers yellowish gray to pale olive. It is well compacted, thin-bedded, poorly-laminated, and usually nonfissile, although occasional well-laminated, moderately fissile layers are present. Numerous lenses and concretions of argillaceous limestone, up to a foot in thickness, occur throughout the mudstone and are locally abundant. The limestone ranges from dark gray to black on fresh surfaces and weathers light gray to pale orange. Fossil mollusks occur in some of the limestone lenses and concretions in the lower 500 ft of the mudstone facies. Poorly preserved tests and molds of foraminifera were also noted at several horizons in the mudstone.

Interbedded with the mudstone are numerous thin beds of hard, fine- to coarse-grained, gray to yellowish brown arkosic sandstone. The sandstone is poorly to well sorted and the individual sand grains are angular to subangular in shape. Many of the sandstone beds are internally massive, but some show well-developed cross-lamination. Small-scale load casts were noted on the basal surfaces of some of the fine-grained beds. The sandstone interbeds are most abundant in the vicinity of tongues of the sandstone facies of the Juncal Formation. As they are traced toward these tongues, they thicken, become coarser in texture and lighter in color, and eventually grade into the typical light gray conglomeratic sandstone of the sandstone facies.

Locally, yellowish gray–weathering sandy siltstone, similar to that of the siltstone facies of the Juncal Formation, is interbedded with the mudstone. The siltstone generally occurs in the vicinity of tongues of the sandstone facies.

Matilija Sandstone

The Matilija Sandstone was originally described by Kerr and Schenck (1928:1090) as a member of the Tejon Formation, but most geologists (e.g., Kelley, 1943; Dibblee, 1950; 1966; Kleinpell and Weaver, 1963; Stauffer, 1967) have subsequently accorded it formation status. Vedder (1972:2) has recently proposed the formal adoption of the Matilija Sandstone as a name of formation rank. The type section of this unit is at Matilija Springs, about 4 mi northwest of Ojai, California, and 10 mi due south of Reyes Peak. At the type section, the formation is 2,500 to 3,000 ft thick and is composed predominantly of sandstone with subordinate interbeds of silty shale and mudstone (T. L. Bailey *in* Redwine, 1952). It lies conformably on the Juncal Formation and is overlain conformably by the Cozy Dell Shale.

Dickinson (1969:16) recognized three distinct lithologic phases within the Matilija Sandstone at the type section: (1) a lower phase ("Ojala Phase") composed of massive, sharply bounded, gray, medium- to coarse-grained, grain-flow sandstone beds lacking megafossils and with prominent mudstone partings between beds; (2) a middle phase ("Reyes Phase") composed of massive and cross-bedded, gray to white, medium- to coarse-grained, locally conglomeratic sandstone containing megafossils indicative of deposition in shallow waters; and (3) an upper phase ("Derrydale Phase") composed of laminated and cross-laminated, green to tan, fine- to medium-grained sandstone.

Strata similar to the middle and upper phases of the Matilija Sandstone occur in the Pine Mountain area and have been referred to this unit by Dickinson (1969:

16–17). In this paper, however, only strata similar to the upper phase of the Matilija Sandstone at its type section are assigned to it. Light gray, coarse-grained, locally conglomeratic sandstone similar to the middle phase of the type Matilija has been referred in this paper to the Juncal Formation because of its lithologic similarity to, and physical continuity with, the sandstone facies of this formation.

The Matilija Sandstone crops out along the flanks of the Piedra Blanca syncline and in the vicinity of the Wegis Ranch, in the southern and western portions of the mapped area. It conformably overlies and intertongues eastward with the sandstone facies of the Juncal Formation and is conformably overlain by the Coldwater Sandstone. Toward the northwest, it intertongues with the Cozy Dell Shale. The formation attains a maximum thickness of about 1,600 ft in the vicinity of Pine Mountain Lodge (see geologic map).

Fine- to medium-grained, laminated and cross-laminated, greenish gray to yellowish brown arkosic sandstone is the principal lithology of the Matilija Sandstone in the Pine Mountain area. The sandstone is thin- to thick-bedded, well sorted, and is composed predominantly of subrounded to subangular grains of quartz and feldspar with small amounts of mica (generally less than 15 percent) and lithic fragments. Dark brown–weathering, ellipsoidal, calcareous sandstone concretions up to 4 ft in diameter are abundant along certain horizons in the sandstone. Frequently, these concretions are associated with lenses of coarse-grained, conglomeratic sandstone, 1 to 4 ft thick, which appear to be of scour-and-fill origin. The basal surfaces of these lenses generally channel into the underlying fine- to medium-grained sandstone beds. Fossil mollusks are abundant in many of the concretions and clay-lined callianassid burrows, similar to those in the sandstone facies of the Juncal Formation, are common in the sandstone.

In the vicinity of Pine Mountain Lodge, an interval of olive gray silty mudstone occurs in the upper 300 ft of the Matilija Sandstone. The mudstone is about 200 ft thick and lithologically resembles the mudstone of the Cozy Dell Shale. Lenses and concretions of dark gray silty limestone, 6 in to 2 ft thick, occur at several horizons in the mudstone and locally contain fossil mollusks.

Cozy Dell Shale

The name Cozy Dell Shale was proposed by Kerr and Schenck (1928:1090) for a unit of rhythmically interbedded silty shale and sandstone, about 2,500 ft thick, which occupies a stratigraphic position between the Matilija Sandstone and the Coldwater Sandstone in the Santa Ynez Range. The type locality is in Cozy Dell Canyon, a minor tributary to the Ventura River north of Ojai and about 10 mi south of Reyes Peak. Kerr and Schenck (ibid., p. 1090) regarded the Cozy Dell Shale as a member of the Tejon Formation. Subsequently, however, most geologists have considered it a formation and Vedder (1972:2) has proposed that the Cozy Dell Shale be formally adopted as a name of formation rank.

Within the mapped area, the Cozy Dell Shale is confined to the Piedra Blanca syncline where it forms a southeastward-lensing tongue within the Matilija Sandstone. It ranges from 0 to about 700 ft in thickness and increases in thickness toward the northwest, beyond the map boundary.

The Cozy Dell Shale is composed predominantly of olive gray to brownish gray, micaceous, silty mudstone with subordinate thin interbeds of gray to yellowish brown, fine- to medium-grained sandstone. The mudstone is thin-bedded to massive and generally nonfissile. The sandstone beds range from 4 in to 2 ft in thickness and are similar in lithology and texture to the Matilija Sandstone. Lenses and concretions of dark gray silty limestone, 6 to 18 in thick, are abundant at certain horizons in the mudstone. Fossil mollusks occur locally in the limestone lenses and concretions and molds of foraminifera were noted at several horizons in the mudstone.

Coldwater Sandstone

The Coldwater Sandstone was first mentioned by Watts (1897) and later described by Kew (1924:26) and Taliaferro (1924:789–802) as the upper member of the Tejon Formation. Subsequently, this unit has generally been considered a formation and Vedder (1972:2) has proposed that the Coldwater Sandstone be formally adopted as a name of formation rank. The type area of the Coldwater Sandstone is in Coldwater Canyon, about 5 mi northwest of Filmore, California, and about 8 mi southeast of Sespe Hot Springs. In its type area, according to Kew (1924:28), the formation lies conformably on older Eocene strata but is unconformable beneath the Sespe Formation. Elsewhere, the upper contact is either a local disconformity or is gradational (Vedder, 1972:2).

Within the mapped area, the Coldwater Sandstone is confined to two isolated outcrops along the axis of the Piedra Blanca syncline, one near Pine Mountain Lodge and the other south of Thorn Point. Only a partial section of the formation, about 700 ft thick, has been preserved. The lower contact of the formation with the Matilija Sandstone is conformable and gradational.

The Coldwater Sandstone in this area is composed predominantly of light gray to pale yellowish green, thick-bedded, cross-stratified, well-sorted arkosic sandstone, with a few thick interbeds of light gray, very coarse-grained, poorly sorted, conglomeratic sandstone similar to that of the sandstone facies of the Juncal Formation. Intraformational conglomerate and scour-and-fill structure are frequently associated with the conglomeratic sandstone. In the vicinity of Pine Mountain Lodge, a thin sequence of brownish green silty mudstone and yellowish brown fine-grained sandstone, about 15 ft thick, occurs about 350 ft above the base of the formation.

BIOSTRATIGRAPHY

Introduction

Fossil mollusks are abundant in certain parts of the Pine Mountain section. Principal mollusk-bearing units are: the lower few hundred feet of the mudstone facies of the Juncal Formation in Hot Springs Canyon; the siltstone and sandstone facies in the upper half of the Juncal Formation; and the Matilija Sandstone. A few species were also collected from the Cozy Dell Shale and the Coldwater Sandstone.

Fossil collections from 78 localities have been examined and a total of 194 molluscan taxa, including 113 gastropods, 78 pelecypods, and 3 scaphopods, have been identified. Besides the mollusks, other fossils identified include one brachiopod, numerous skeletal fragments of callianassid and brachyurian crustaceans, and several species of foraminifera. The stratigraphic distribution of the mollusks and other fossils in the Pine Mountain section is summarized in table 1 (in pocket).

The preservation of the mollusks is generally poor, although exceptionally well-preserved specimens were obtained from some localities. Most of the mollusks occur in hard calcareous concretions and are difficult to extract without breakage. At certain horizons, particularly in the lower part of the mudstone facies of the Juncal Formation and in the mudstone beds near the top of the Matilija Sandstone, many of the shells are deformed. This adds to the problem of identification.

Four stratigraphically distinct molluscan assemblages are recognized in the Pine Mountain section. For convenience, these assemblages are informally designated the *Turritella uvasana infera, Turritella uvasana applinae, Ectinochilus supraplicatus,* and *Ectinochilus canalifer* faunas. The first three faunas occur within the Juncal Formation; the *Ectinochilus canalifer* fauna includes the mollusks in the Matilija, Cozy Dell and Coldwater formations and in the upper 1,000 ft of the Juncal Formation. These faunas are closely related and overlap in composition but each contains some taxa that are restricted to it. The stratigraphic positions of these faunas in the Pine Mountain section are indicated in figure 2.

Turritella uvasana infera FAUNA

The *Turritella uvasana infera* fauna characterizes about 375 ft of strata in the lower part of the mudstone facies of the Juncal Formation in Hot Springs Canyon (geologic map; fig. 3). The lowest occurrence of the fauna is in a gray calcareous sandstone bed, 3 to 5 ft thick, which is about 50 ft stratigraphically above the base of the mudstone facies and 150 ft above the base of the Juncal Formation. The fossils occur mainly within thin, fine-grained, sandstone beds interbedded with the mudstone. A few species also occur free in the mudstone and in fine-grained limestone concretions and lenses.

The *Turritella uvasana infera* fauna is characterized by the lowest occurrence in the Pine Mountain section of *Ectinochilus macilentus, Turricula praeattenuata, Surculites mathewsonii, Sassia bilineata, Globularia hannibali, Akera maga, Sinum obliquum, Ficopsis remondii crescentensis, Cylichnina tantilla, Calyptraea diegoana, Nuculana parkei, Corbula parilis* and *Acila decisa;* and the restricted occurrence in this section of many taxa, including *Velates perversus, Turritella andersoni* s.s., *T. uvasana infera, T. buwaldana crooki, Tibia llajasensis, Chedevillea stewarti, Galeodea sutterensis, Pachycrommium?* n. sp., *Siphonalia* cf. *S. thunani, Falsifusus* cf. *F. marysvillensis, Eosurcula capayana, Cryptochorda* cf. *C. californica, Crommium andersoni, Clavilithes tabulatus, Cerithium excelsum, Cerithium* cf. *C. orovillensis, Campanilopa dilloni, Brachysphingus mammilatus, Ampullella hewitti, Venericardia hornii lutmani, Porterius woodfordi, Pholadomya* n. sp., *Odontogryphaea? haleyi, Nayadina llajasensis, Nuculana hondana, Ledina fresnoensis, Arca* n.sp.?, *Glycymeris* aff. *G. fresnoensis, Callocardia conradi, Pitar* cf. *P. palmeri,* and *Limopsis marysvillensis.*

Turritella andersoni s.s. and *T. uvasana infera* are the most abundant taxa

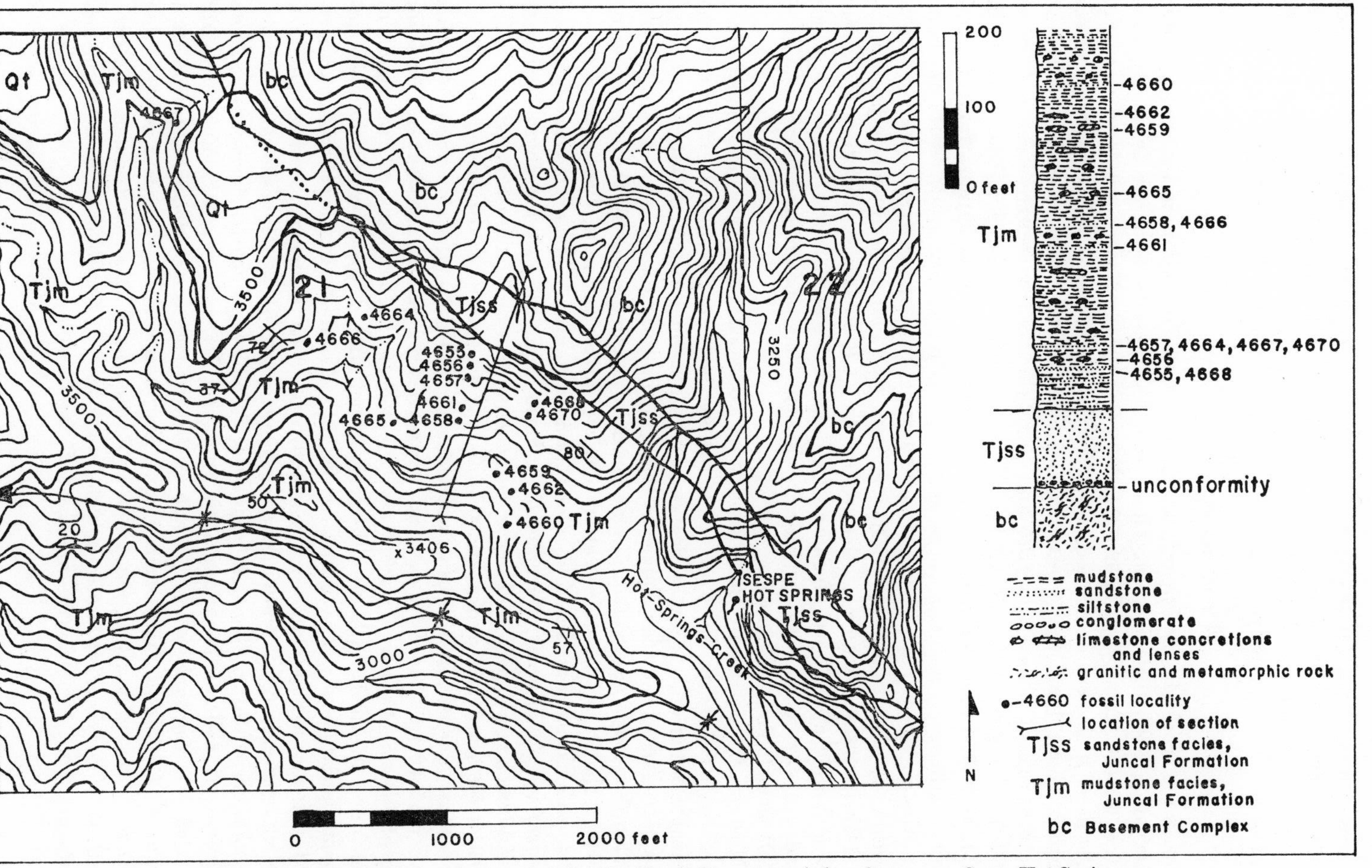

Fig. 3. Detailed map of the strata containing the *Turritella uvasana infera* fauna near Sespe Hot Springs.

in this fauna. Thin sandstone beds composed of hundreds of current-oriented shells of these subspecies occur at several horizons in the mudstone. *Velates perversus* and *Ampullella hewitti* are abundant at UCR locality 4668. Many of the taxa in this fauna, however, are represented by only one or a few specimens.

Turritella uvasana applinae FAUNA

This fauna occurs in the upper half of the Juncal Formation, in the stratigraphic interval between the mudstone facies and the uppermost tongue of the siltstone facies. In the vicinity of Thorn Meadows and Grade Valley (geologic map; fig. 4), where it is best developed, the *Turritella uvasana applinae* fauna ranges through about 1,500 ft of section. The lowest stratigraphic occurrences of the fauna are at UCR localities 4673 and 4751; the highest occurrence is at locality 4690. In the San Guillermo syncline, UCR localities 4694, 4699, 4700, 4701, 4702, 4703, and 4750 also contain this fauna. Because of facies changes, faulting, and the sporadic distribution of the fossils, the relative stratigraphic positions of the localities in the San Guillermo syncline and those in the Grade Valley–Thorn Meadows area are uncertain. UCR localities 4699 and 4700 appear to be approximately in the same part of the section as UCR localities 4751 and 4673 near Grade Valley and Thorn Meadows; UCR localities 4694 and 4703, on the other hand, appear to be stratigraphically higher than the highest locality (UCR 4690) near Thorn Meadows.

Most of the fossils in this fauna were collected from limestone concretions associated with thin, fine-grained, sandstone beds in the siltstone facies of the Juncal Formation. A few collections were obtained from coarse-grained, conglomeratic, calcareous sandstone in the sandstone facies.

The *Turritella uvasana applinae* fauna is characterized by the lowest occurrence in the Pine Mountain section of *Laevityphis antiquus, Coalingodea tuberculiformis, Galeodea susanae, Cerithium cliffensis, Pachycrommium? clarki, Pleurofusia fresnoensis, Pyramidella etheringtoni, Strepsidura ficus, Turritella buwaldana* s.s., *Euspira nuciformis, Olivella mathewsonii, Acanthocardia brewerii, Crassatella uvasana semidentata, Glyptoactis domenginica, Glycymeris sagittata, G. rosecanyonensis,* and *Tellina soledadensis;* and the highest occurrence in this section of *Akera maga, Surculites mathewsonii, Ectinochilus macilentus, Turricula praeattenuata, Nuculana parkei,* and *Acila decisa.* Many taxa are restricted to this fauna in the Pine Mountain section, including *Ficopsis cooperiana, Homalopoma umpquaensis, Architectonica cognata, A. ullreyana, Bonellitia paucivaricata, Fusinus teglandae, Euspira clementensis, Natica rosensis, Polinices gesteri, Proximitra? cretacea, Pseudoperissolax blakei praeblakei, Olequahia domenginica, Scaphander costatus, Solariella crenulata, Tejonia lajollaensis, Turritella uvasana applinae, T. andersoni lawsoni, T. scrippsensis, Xenophora stocki, Volutocristata lajollaensis, Barbatia morsei, Claibornites diegoensis, Diplodonta unisulcatus, Linga* cf. *L. taffana, Isognomon n.* sp.?, *Gari eoundulata, Pelecyora gabbi, Pitar kelloggi, P. avenalensis, P. joaquinensis, Periploma* cf. *P. stewartvillensis, Callista* cf. *C. tecolotensis, Portlandia rosa, Solena coosensis, S. subverticala, Thracia sorrentoensis,* and *Venericardia hornii* cf. *V. hornii calafia.*

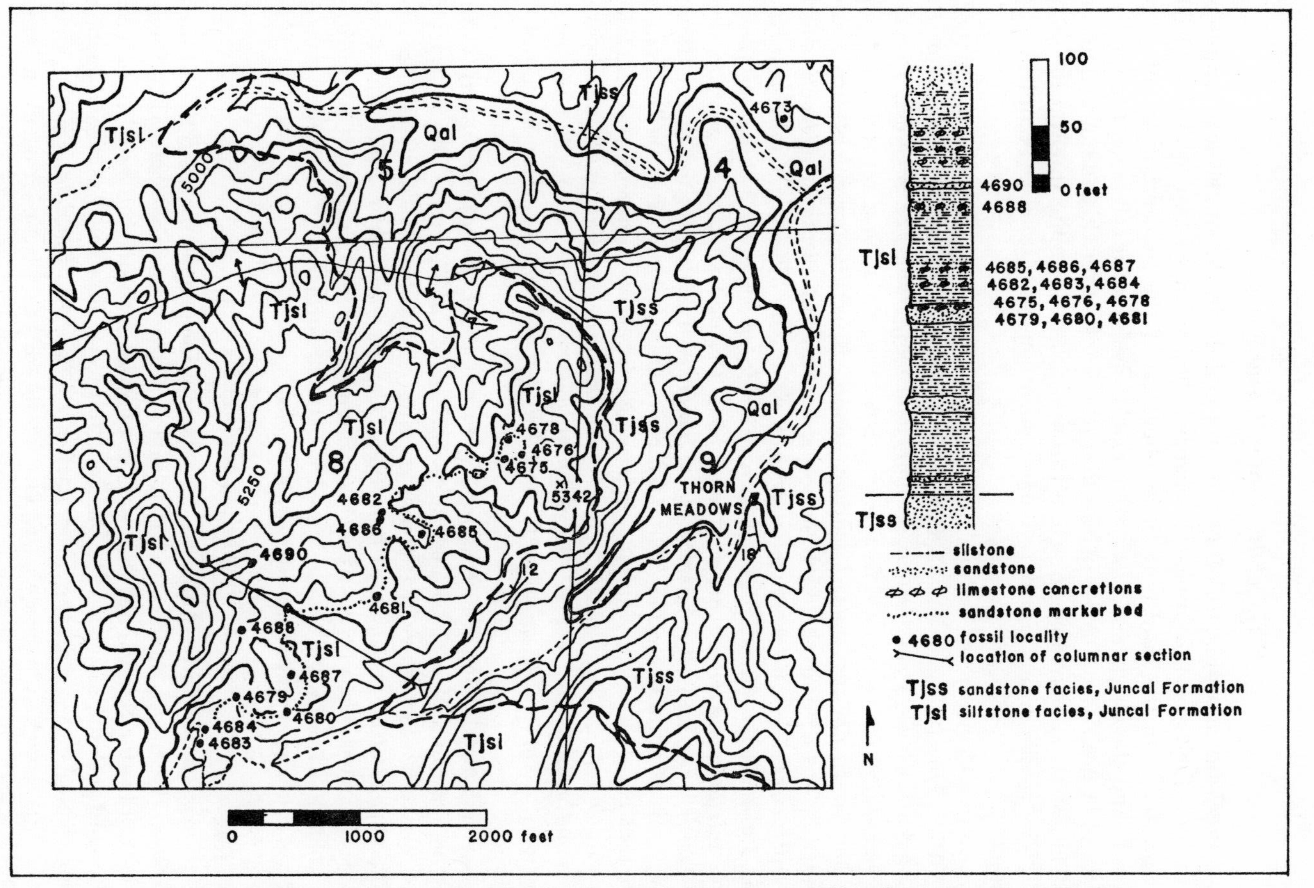

Fig. 4. Detailed map of the strata containing the *Turritella uvasana applinae* fauna near Thorn Meadows.

Ectinochilus supraplicatus Fauna

The *Ectinochilus supraplicatus* fauna occurs mainly in the uppermost tongue of the siltstone facies of the Juncal Formation, but includes two localities (UCR 4695 and 4698) in the sandstone facies. Because of the sporadic distribution of the fossils and the absence of distinctive marker beds in the siltstone, the relative stratigraphic positions of most of the localities containing this fauna are uncertain. The fauna is estimated to range through a stratigraphic interval of about 600 ft. The fossils occur in concretionary calcareous sandstone beds and in silty mudstone.

The *Ectinochilus supraplicatus* fauna is characterized by the lowest occurrence in the Pine Mountain section of *Turritella uvasana neopleura, Nekewis io, Ficopsis hornii, Terebra californica, Tejonia moragai, Ranellina pilsbryi, Pseudoliva inornata, Natica uvasana, Loxotrema turritum, Potamides carbonicola, Polinices hornii, Pitar uvasanus, Pachecoa hornii, Brachidontes cowlitzensis,* and *Cyclinella elevata;* and the highest occurrence in this section of *Coalingodea tuberculiformis, Ficopsis remondii crescentensis, Sassia bilineata, Globularia hannibali, Pyramidella etheringtoni, Pleurofusia fresnoensis, Cerithium cliffensis, Tellina soledadensis, Glyptoactis domenginica, Corbula parilis, Crassatella uvasana semidentata,* and *Glycymeris rosecanyonensis.* Taxa restricted to this fauna in the Pine Mountain section include *Exilia microptygma, Turbonilla gesteri, Pseudoliva volutaeformis, Neverita globosa, Ectinochilus supraplicatus, Molopophorus antiquatus, Conus* n. sp.? aff. *C. californianus, Scalina* cf. *S. aragoensis, Perse martinez, Conomitra* aff. *C. washingtoniana, Tivelina* cf. *T. vaderensis, Glycymeris* n. sp.? aff. *G. perrini, Pitar soledadensis, Corbula dickersoni,* and *C. torreyensis.*

Two geographic facies are recognizable within the *Ectinochilus supraplicatus* fauna: a western facies (UCR localities 4705, 4706, 4707, and 4753) characterized by a diverse assemblage of typically marine genera such as *Turritella, Pseudoliva, Terebra, Polinices, Natica, Pleurofusia, Conus, Globularia, Pitar,* and *Crassatella;* and an eastern, probably brackish water, oyster bank facies (UCR localities 4695, 4696, 4697, and 4698) dominated by *Ostrea idriaensis* and the gastropods *Loxotrema turritum* and *Potamides carbonicola.*

Ectinochilus canalifer Fauna

The *Ectinochilus canalifer* fauna includes the molluscan assemblage in the Matilija Sandstone, the upper 1,000 ft of the Juncal Formation, the Cozy Dell Shale, and the Coldwater Sandstone. It is best developed in the vicinity of Pine Mountain Lodge (geologic map; fig. 5), where it characterizes about 2,000 ft of section. It is also well developed in the lower part of the Matilija Sandstone in the vicinity of Wegis Ranch (geologic map). The fossils for the most part occur in large, dark brown-weathering, calcareous sandstone concretions, except in the mudstone beds near the top of the Matilija Sandstone and in the Cozy Dell Shale where they occur in fine-grained, light gray to brown–weathering, limestone concretions.

The *Ectinochilus canalifer* fauna is characterized by the highest occurrence in the Pine Mountain section of *Turritella uvasana neopleura, T. buwaldana* s.s.,

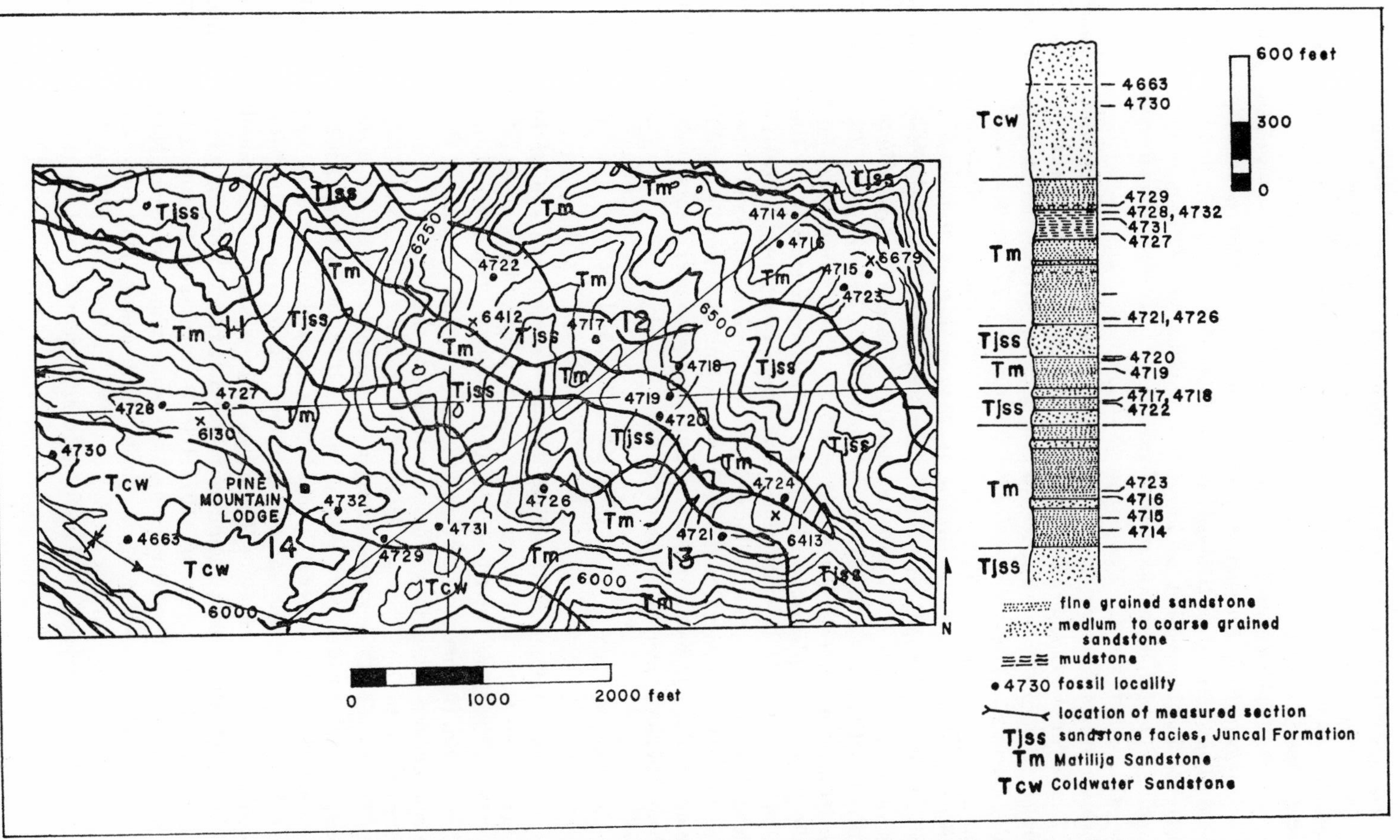

Fig. 5. Detailed map of the strata containing the *Ectinochilus canalifer* fauna near Pine Mountain Lodge.

Nekewis io, Turricula cohni, Terebra californica, Polinices hornii, Natica uvasana, Euspira nuciformis, Tejonia moragai, Strepsidura ficus, Pachycrommium? clarki, Potamides carbonicola, Loxotrema turritum, Sinum obliquum, Pseudoliva inornata, Ranellina pilsbryi, Olivella mathewsonii, Ficopsis hornii, Laevityphis antiquus, Cylichnina tantilla, Calyptraea diegoana, Pachecoa hornii, Ostrea idriaensis, Nemocardium linteum, Glycymeris sagittata, Pitar uvasanus, Cyclinella elevata, Brachidontes cowlitzensis, and *Acanthocardia brewerii;* and the restricted occurrence in this section of *Turritella uvasana* s.s., *T. uvasana sargeanti, T. schencki, Siphonalia sopenahensis, Neverita secta, Hexaplex? whitneyi, Perse sinuata, Olequahia hornii, Molopophorus tejonensis, "Trichotropis" lajollaensis, Hemipleurotoma pulchra, Gemmula abacta, Ficus mamillata, Ficopsis remondii* s.s., *Exilia fausta, Ectinochilus canalifer, Crepidula pileum, Conus hornii, Acteon quercus, Abderospira hornii, Nerita triangulata, Architectonica hornii, Venericardia hornii* s.s., *Thracia dilleri, Periploma* n. sp.?, *Tellina tehachapi, T. lebecki, T.* aff. *T. townsendensis, T. castacana, Macoma viticola, Spisula bisculpturata, Pteria pellucida, Pitar californianus, P. dickersoni, P.* cf. *P. lascrucensis, Callista hornii, C. conradiana, C. andersoni, Nuculana gabbi, N.* cf. *N. washingtonensis, Glycymeris viticola, Gari hornii, Crassatella uvasana* s.s., and *Corbula hornii.*

In the section near Pine Mountain Lodge, three subspecies of *Turritella uvasana* are stratigraphically restricted within the *Ectinochilus canalifer* fauna. *Turritella uvasana neopleura,* which has its lowest occurrence in the Pine Mountain section within the beds containing the *Ectinochilus supraplicatus* fauna, has its highest occurrence within the lower 200 ft of the Matilija Sandstone; *T. uvasana* s.s. occurs in the stratigraphic interval from about 200 ft to about 800 ft above the base of the Matilija Sandstone; and *T. uvasana sargeanti* occurs from about 1,000 ft to about 1,600 ft above the base of the unit.

AGE AND CORRELATION

Introduction

The molluscan faunas in the Pine Mountain section are similar in taxonomic composition to those that characterize the "Capay," "Domengine," "Transition," and "Tejon" "Stages" of Clark and Vokes (1936). The sequence of the faunas in this section, furthermore, confirms the chronological sequence of these "Stages." The relationship of the Pine Mountain faunas to the "Stages," and to the various zones assigned to these "Stages" by previous authors, is illustrated in figure 6.

On the basis of indirect correlations with Europe, the "Capay" through "Tejon" "Stages" were considered by Weaver et al. (1944, pl. 1) to represent all but the latest portion of the Eocene Epoch. The "Capay Stage" was regarded as early Eocene, the "Domengine" and "Transition" "Stages" as middle Eocene, and the "Tejon Stage" as late Eocene. These age assignments are provisionally accepted in this report, although it should be noted that recent work by planktonic foraminiferal specialists (Schmidt, 1971:190; Steineck and Gibson, 1971:478–479) suggests that the "Tejon Stage" and the approximately coeval Narizian Stage (Mallory, 1959) of benthonic foraminiferal chronology may be of middle rather than late Eocene age. A middle Eocene age for the "Tejon Stage" is also suggested by the occurrence of planktonic foraminifera and calcareous nannoplank-

SERIES	"STAGES" (Clark & Vokes, 1936, Weaver et al., 1944)	ZONES		PINE MT. FAUNAS
		Clark & Vokes (1936)	Weaver et al. (1944)	
EOCENE	"TEJON"	Turritella uvasana sargeanti	Turritella olequahensis	Ectinochilus canalifer
			Turritella sargeanti	
		Turritella uvasana uvasana	Turritella uvasana s.s.	
	"TRANSITION"	Rimella supraplicata	Turritella applini	Ectinochilus supraplicatus
	"DOMENGINE"	Galeodea tuberculiformis	Turritella lawsoni	Turritella uvasana applinae
	"CAPAY"	Galeodea susanae	Galeodea susanae	
		Siphonalia sutterensis	Turritella andersoni	Turritella uvasana infera

Fig. 6. Relationship of the Pine Mountain faunas to the Eocene "Stages" and "Zones" of Clark and Vokes (1936) and Weaver et al. (1944).

ton characteristic of the late Eocene (Bartonian Stage) in the superjacent Refugian Stage (Lipps, 1967; Brabb et al., 1971). If the "Tejon Stage" is of middle Eocene age, then some of the other Eocene "Stages" may also be older than previously thought.

Turritella uvasana infera Fauna

The *Turritella uvasana infera* fauna contains several taxa, including *Galeodea sutterensis, Clavilithes tabulatus, Turritella andersoni* s.s., and *T. buwaldana crooki,* which previous authors (Clark and Vokes, 1936:860, 861; Merriam and Turner, 1937:92; Vokes, 1939:33, 35, 36) have regarded as index fossils for the "Capay Stage." Several other taxa that are restricted to this fauna in the Pine Mountain section have also been recorded elsewhere on the Pacific Coast only from strata assigned to the "Capay Stage." *Limopsis marysvillensis* has been previously reported only from its type locality in Dickerson's (1913; 1916) *Siphonalia sutterensis* Zone in the Marysville Claystone Member of the Meganos Formation (Stewart, 1949) at Marysville Buttes in the Sacramento Valley, California, and from the type locality of the "Capay Stage" in Capay Valley (Merriam and Turner, 1937:94, 99). *Eosurcula capayana* and *Nuculana hondana* have been previously reported (Vokes, 1939:29, 30, 35) only from their type locality in the *Turritella andersoni*-bearing glauconitic sandstone bed near the base of the Cerros Shale Member of the Lodo Formation (White, 1938) (= the Lower Zone of the Ragged Valley Shale Member of the Arroyo Hondo Formation of Vokes,

1939) at Salt Creek, north of Coalinga, California. *Chedevillea stewarti* and *Tibia llajasensis* were described by Clark (1942:116) from fine silts in the lower part of the Llajas Formation on the south side of the Simi Valley in Ventura County, California. The lower part of the Llajas Formation has generally been considered to be of "Capay" age (Clark and Vokes, 1936:856, 860, fig. 2; Merriam and Turner, 1937:92, 98; Weaver et al., 1944:577, pl. 1). *Chedevillea stewarti* has also been questionably identified, along with *Galeodea* cf. *G. sutterensis* and *Clavilithes* cf. *C. tabulatus*, from the lower part of the Maniobra Formation in eastern Riverside County, California (Crowell and Susuki, 1959:588–590, pl. 2). The lower part of this unit was assigned by Crowell and Susuki to the "Capay Stage." *Campanilopa dilloni* and *Ampullella hewitti* have been previously reported (Hanna and Hertlein, 1949) only from the unnamed sandstone member in the middle of the Lodo Formation near Media Agua Creek in Kern County, California. Mallory (1959:34, 75, 76) considered this sandstone member, which at Media Agua Creek contains his type Upper Penutian *Alabamina wilcoxensis* Zone foraminiferal assemblage, to be of early Eocene ("Capay") age.

Additional evidence for the "Capay" age of the *Turritella uvasana infera* fauna is the overlap or joint occurrence of taxa that have been reported (Anderson and Hanna, 1925; Hanna, 1927; Clark and Vokes, 1936; Merriam and Turner, 1937; Turner, 1938; Vokes, 1939; Stewart, 1946; 1949; Weaver, 1949; 1953) elsewhere on the Pacific Coast only from "Capay" and younger Eocene strata, including *Velates perversus, Globularia hannibali, Ficopsis remondii crescentensis, Crommium andersoni, Surculites mathewsonii, Turricula praeattenuata, Sassia bilineata, Sinum obliquum, Ectinochilus macilentus, Cylichnina tantilla, Porterius woodfordi, Nuculana parkei,* and *Corbula parilis;* with taxa that have been reported (Clark and Woodford, 1927; Turner, 1938; Vokes, 1939; Merriam, 1941) elsewhere only from "Capay" and older ("Meganos") strata, including *Turritella uvasana infera, Venericardia hornii lutmani,* and *Callocardia conradi. Akera maga,* which ranges up into the "Domengine" *Turritella uvasana applinae* fauna in the Pine Mountain section, has been previously reported (Vokes, 1939) only from the "Domengine Stage." Its occurrence in the *Turritella uvasana infera* fauna is considered to extend its range down into the "Capay Stage."

One species in the *Turritella uvasana infera* fauna, *Brachysphingus mammilatus,* has been previously regarded as an index fossil for the "Meganos Stage" (Clark and Vokes, 1936:858; Merriam and Turner, 1937:96). The joint occurrence of this species in the Pine Mountain section with many species that elsewhere have not been reported to occur in strata older than the "Capay Stage" is here taken as evidence that its range extends into this "Stage." Merriam and Turner (1937:97) also reported the occurrence of a *Brachysphingus* very similar to *B. mammilatus* in association with species characteristic of the "Capay Stage" in Round Valley in the northern Coast Ranges of California.

Schlee (1952:71, 72) reported the occurrence of the "Meganos" index species *Turritella meganosensis* at one of the localities (UCR 4664) in the lower part of the mudstone facies of the Juncal Formation and, on this basis, assigned some of these strata to the "Meganos Stage." No specimens of *Turritella me-*

ganosensis were obtained from locality 4664 during this study. Large specimens of *Turritella andersoni* s.s., however, are very abundant at this locality, although they were not reported by Schlee. Many of the specimens are deformed and it is possible that they were mistaken by Schlee for *Turritella meganosensis*.

In the vicinity of Coalinga, California, Clark and Vokes (1936:860; see also Vokes, 1939:29–36) recognized two faunal zones in strata that they assigned to the "Capay Stage": a lower zone characterized by *Galeodea sutterensis* and containing a fauna similar to that of Dickerson's (1913; 1916) *Siphonalia sutterensis* Zone at Marysville Buttes and to that of the Capay Formation in Capay Valley, type locality of the "Capay Stage"; and an upper zone characterized by *Galeodea susanae* and containing a fauna similar to that of the overlying type Domengine Formation, type locality of the "Domengine Stage." The upper faunal zone was referred by Clark and Vokes (1936:860) and Vokes (1939:36) to the "Capay Stage" because of its presumably similar stratigraphic position to the upper part of the type Capay Formation, although the authors noted that the upper part of this formation had not yielded a molluscan fauna. The *Turritella uvasana infera* fauna contains species characteristic of only the lower of Clark and Vokes's two faunal zones of the "Capay Stage." *Galeodea susanae,* the index fossil for the upper zone, occurs in the *Turritella uvasana applinae* fauna (and has also been doubtfully identified in the *Ectinochilus supraplicatus* fauna) where it is associated with species characteristic of the "Domengine Stage."

Turritella uvasana applinae FAUNA

The *Turritella uvasana applinae* fauna contains taxa characteristic of the "Domengine Stage" and the upper part of the "Capay Stage." The following taxa in this fauna have been reported (Vokes, 1939; Stewart, 1946; Weaver, 1949; 1953) elsewhere on the Pacific Coast only from strata assigned to the "Domengine Stage": *Turritella andersoni lawsoni, Olequahia domenginica, Proximitra? cretacea, Pitar joaquinensis, P. avenalensis,* and *Pelecyora gabbi.* The first four taxa have been regarded as index fossils for the "Domengine Stage" (Vokes, 1939:33–35; Weaver et al., 1944, pl. 1). *Coalingodea tuberculiformis* and *Pleurofusia fresnoensis* have also been previously regarded as index fossils for the "Domengine Stage" (Clark and Vokes, 1936:861, fig. 1; Vokes, 1939:33). In the Pine Mountain section, however, these species range up into the "Transition Stage."

Additional evidence for the "Domengine" age of the *Turritella uvasana applinae* fauna is the joint occurrence of taxa that elsewhere on the Pacific Coast have been reported (Stewart, 1926:387, 388; Hanna, 1927; Clark, 1938; Vokes, 1939; Merriam, 1941; Weaver, 1953) only from "Domengine" and younger ("Tejon") Eocene strata, including *Euspira clementensis, Laevityphis antiquus, Turritella buwaldana* s.s., and *Crassatella mulates;* with taxa that have been reported (Clark and Woodford, 1927; Hanna, 1927; Merriam and Turner, 1937; Turner, 1938; Vokes, 1939; Stewart, 1946; Weaver, 1953) elsewhere only from "Domengine" and older Eocene ("Capay") or late Paleocene ("Meganos") strata, including *Turricula praeattenuata, Ectinochilus macilentus, Polinices gesteri, Architectonica cognata, Akera maga, Surculites mathewsonii,* and *Gari eoundulata.* Several other taxa that have their lowest occurrence in the *Turritella uvasana applinae* fauna in the Pine Mountain section, including *Cerithium cliffensis, Pyramidella etheringtoni, Strepsidura ficus, Glyptoactis domenginica,* and *Glycymeris rosecanyonensis,* also have not been reported from strata older than the "Domengine Stage" elsewhere on the Pacific Coast. One species, *Architectonica ullreyana,* has previously been reported (Dickerson, 1916:487) only from strata assigned to the "Capay Stage" (viz., the *Siphonalia sutterensis* Zone at Marysville Buttes). Its occurrence in the *Turritella uvasana applinae* fauna extends its range into the "Domengine Stage."

Of particular significance in the *Turritella uvasana applinae* fauna is the presence of *Galeodea susanae* Schenck. This species has previously been regarded as an index fossil for the "Upper Capay Stage" *Galeodea susanae* Zone (Clark and Vokes, 1936:860, fig. 1; Vokes, 1939:33; Weaver et al., 1944, pl. 1). This Zone at its type locality in the upper part of the Lodo Formation near Coalinga, California, contains a fauna very similar to that of the overlying Domengine Formation, the type of the "Domengine Stage." Vokes (1939:31) noted that of the 27 fully determinable species in the *Galeodea susanae* Zone, all but four also occur in the Domengine Formation. The *Galeodea susanae* Zone was distinguished from the "Domengine Stage" primarily on the assumption that *Galeodea susanae* was restricted to stratigraphic horizons below those characterized by *Coalingodea tuberculiformis* (Hanna), which was considered to be an index fossil for the "Domengine Stage" (Vokes, 1939:33). In the Pine Mountain section, however, these species are directly associated (UCR locality 4683) in the *Turritella uvasana applinae* fauna. They have also been reported to occur together in the Llajas Formation in the Simi Valley, Ventura County, California (Schenck, 1926:83–85, California Academy of Sciences fossil locality no. 364). Thus, their stratigraphic ranges overlap.

Although they recognized the close relationship of the faunas of the *Galeodea susanae* Zone and the "Domengine Stage," Clark and Vokes (1936:860) and Vokes (1939:32, 36) assigned the *G. susanae* Zone to the "Capay Stage" because of its presumably similar stratigraphic position to the upper part of the type Capay Formation in Capay Valley, California. This assignment is questionable because evidence for the "Capay" age of the upper part of the type Capay Formation is lacking. Merriam and Turner (1937:94) stated that the typical "Capay" fauna was collected from two closely spaced stratigraphic horizons in the lower 500 ft of the Capay Formation, which is more than 2,000 ft thick at the type section (Crook and Kirby, 1935:334, 335). Strata in the upper part of the type Capay Formation have not yielded a molluscan fauna and may be younger than "Capay" age.

Because the *Galeodea susanae* Zone is faunally almost indistinguishable from the "Domengine Stage," because the basis on which it was distinguished from that "Stage" (viz., the assumption that *Galeodea susanae* is restricted to stratigraphic horizons below those characterized by *Coalingodea tuberculiformis*) is invalid, and because unequivocal evidence for its "Capay" age is lacking, it is here considered to be biostratigraphically and chronostratigraphically equivalent to the "Domengine Stage" and is reassigned to that "Stage." The "Capay Stage," accordingly, is restricted to include only the lower of the two faunal zones assigned to it by Clark and Vokes (1936) and others (Vokes, 1939; Weaver et al., 1944). It follows from this revision that the base of the "Domengine Stage" in the type section does not coincide with the base of the Domengine Formation, as supposed by Clark and Vokes (1936) and Vokes (1939), but occurs at an undetermined horizon within the underlying Lodo Formation.

In addition to *Galeodea susanae,* several other taxa in the *Turritella uvasana applinae* fauna have previously been considered to be characteristic of strata younger or older than the "Domengine Stage." *Turritella uvasana applinae* and *T. scrippsensis* have been regarded by some authors (Weaver et al., 1944, pl. 1; Kleinpell and Weaver, 1963:106) as index fossils for the "Transition Stage." *Homalopoma umpquaensis,* on the other hand, was considered by Merriam and Turner (1937:92) to be restricted to the "Capay Stage." The occurrence of these taxa in the Pine Mountain section stratigraphically between molluscan faunas referable to the "Capay" and "Transition" "Stages" and their association with taxa characteristic of the "Domengine Stage" indicate that their ranges extend into this "Stage." *Turritella uvasana applinae* is probably confined to the "Domengine Stage" because it has an apparent ancestor, *T. uvasana infera,* in the "Capay Stage" and an apparent descendant, *T. uvasana neopleura,* in the "Transition Stage."

The *Turritella uvasana applinae* fauna shares more taxa in common with the Rose Canyon Shale Member of the La Jolla Formation (Hanna, 1926; 1927) near San Diego, California, than with any other unit on the Pacific Coast (except, perhaps, the Llajas Formation, the fauna of which is largely undescribed). Of the 67 fully identified species and subspecies in this fauna, 40 (or 60 percent) also occur in the Rose Canyon Shale (Hanna, 1927). Many of the taxa that are characteristically associated in the *Turritella uvasana applinae* fauna in the Pine Mountain section, including *Xenophora stocki, Volutocristata lajollaensis, Turritella uvasana applinae, T. scrippsensis, Turricula praeattenuata, Tejonia lajollaensis, Solariella crenulata, Pyramidella etheringtoni, Natica rosensis, Euspira clementensis, Fusinus teglandae, Ficopsis cooperiana, Cerithium cliffensis, Thracia sorrentoensis, Portlandia rosa, Pitar kelloggi, Glycymeris rosecanyonensis, Crassatella mulates, Clairbornites diegoensis,* and *Cardiomya israelskyi,* were described from the Rose Canyon Shale.

The La Jolla Formation was originally correlated by Hanna (1927:262, 263) with the type Domengine Formation. Later, however, Clark and Vokes (1936:863) suggested that the upper part of the Rose Canyon Shale Member may be of "Transition" age based upon the reported occurrence of *Nekewis io* (Gabb) in these strata. Clark (1943:188) subsequently assigned all of the Rose Canyon Shale to the "Transition Stage" and designated it as the type section, but did not cite any faunal or stratigraphic evidence in support of a "Transition" age for this unit. The close similarity between the *Turritella uvasana applinae* fauna and the fauna of the Rose Canyon Shale leaves little doubt that they are at least partly time-equivalent and that the Rose Canyon Shale is in part of "Domengine" age. This conclusion is supported by the occurrence of the "Domengine" index fossil, *Turritella andersoni lawsoni,* in the Rose Canyon Shale

(Hanna, 1927). This unit, therefore, is unsatisfactory as a type section for the "Transition Stage."

The classification and nomenclature of the Eocene strata near San Diego, California, have recently been revised by Kennedy and Moore (1971). These authors elevated the La Jolla Formation to group rank and raised Hanna's (1926) two lower members (the Delmar Sand and Torrey Sand) of the unit to formational status, but abandoned the name Rose Canyon Shale for a subdivision of the La Jolla Group. They recognized four new formations in the strata formerly mapped by Hanna (1926) as Rose Canyon Shale. From oldest to youngest, these formations are: Mount Soledad Formation, a marine cobble-conglomerate and sandstone unit; Ardath Shale, a richly fossiliferous marine silty shale; Scripps Formation, a moderately fossiliferous marine sandstone unit; and Friars Formation, a chiefly nonmarine sandstone unit. Most of the molluscan taxa described by Hanna (1927) from the Rose Canyon Shale occur in the strata assigned by Kennedy and Moore to the Ardath and Scripps formations (Givens, unpub. data). Kennedy and Moore (1971:716, 717) considered the Ardath Shale to be of middle Eocene age and the Scripps Formation of middle- and late-Eocene age. They reported the following mollusks from the type section of the Ardath Shale: *Turritella uvasana applinae, Ficopsis cooperiana,* and *Tejonia lajollaensis.* These taxa and *Turritella andersoni lawsoni* were also reported from the basal strata of the Scripps Formation near its type section. All of these taxa are restricted to the *Turritella uvasana applinae* fauna in the Pine Mountain section and are believed to be restricted to the "Domengine Stage." Kennedy and Moore (1971:717) reported the joint occurrence of *Nekewis io* and *Ectinochilus supraplicatus* in the upper part of the Scripps Formation. As discussed below, the joint occurrence of these species is characteristic of the "Transition Stage." In summary, faunal evidence suggests that the Ardath Shale and the lower part of the Scripps Formation are correlative with the strata containing the *Turritella uvasana applinae* fauna in the Pine Mountain section and are of "Domengine" age. The upper part of the Scripps Formation, including strata formerly assigned to the upper part of the Rose Canyon Shale, is referable to the "Transition Stage" and is correlative with the beds containing the *Ectinochilus supraplicatus* fauna in the Pine Mountain section.

Ectinochilus supraplicatus FAUNA

The *Ectinochilus supraplicatus* fauna is referable to the "Transition Stage." This "Stage" was tentatively proposed by Clark and Vokes (1936:863) for "a distinct faunal group, which includes elements that have heretofore been thought to be confined to the Domengine and lower faunas, associated with forms considered to be diagnostic of the Tejon Stage." Clark and Vokes did not explicitly describe the fauna of the "Transition Stage." They did, however, refer to a molluscan fauna collected by Dreyer (1935) from strata exposed along the Pine Mountain fault (and within the present study area) which contained three species whose association was considered to be diagnostic of this "Stage:" *Nekewis io* (Gabb), *Globularia hannibali* (Dickerson), and *Ectinochilus suprap-*

licatus (Gabb). The first species previously had been known only from the "Tejon Stage," the second species had not been reported previously above the "Domengine Stage," and the third species was considered by Clark and Vokes to be one of the most characteristic species of the "Transition Stage."

The *Ectinochilus supraplicatus* fauna occurs in the same part of the Eocene section in the Pine Mountain area from which Dreyer's (1935) fauna was collected and contains the three species, *Nekewis io, Globularia hannibali,* and *Ectinochilus supraplicatus,* whose joint occurrence was considered by Clark and Vokes to be characteristic of the "Transition Stage." This fauna, furthermore, corresponds precisely in stratigraphic position and taxonomic composition with Clark and Vokes's concept of the "Transition Stage." Stratigraphically, the *Ectinochilus supraplicatus* fauna occurs between faunas characteristic of the "Domengine" and "Tejon" "Stages." Taxonomically, it is characterized by a unique association of taxa that previously have not been reported to occur above the "Domengine Stage," including *Coalingodea tuberculiformis, Sassia bilineata, Pyramidella etheringtoni, Pleurofusia fresnoensis, Perse martinez, Molopophorus antiquatus, Ficopsis remondii crescentensis, Cerithium cliffensis, Tellina soledadensis, Crassatella uvasana semidentata, Glyptoactis domenginica,* and *Corbula parilis;* and taxa that previously have not been reported to occur below the "Tejon Stage," including *Natica uvasana, Pseudoliva inornata, P. volutaeformis, Tejonia moragai, Turricula cohni, Ficopsis hornii, Exilia microptygma,* and *Pitar uvasanus* s.s. Thus, the "Transition" age of the *Ectinochilus supraplicatus* fauna is well established.

Besides the Pine Mountain area, only two other sections are presently known to contain molluscan assemblages referable to the "Transition Stage": the upper part of the Scripps Formation near San Diego, California; and the lower part of the Matilija Sandstone in the Santa Ynez Range west of Santa Barbara, California. The evidence for the "Transition" age of the upper part of the Scripps Formation has already been discussed. Kelley (1943:9) reported the following taxa from the basal Matilija Sandstone near Santa Anita Canyon west of Santa Barbara.

Gari hornii (Gabb)
Macrocallista hornii (Gabb)
Microcallista tecolotensis (Hanna)
Nemocardium linteum (Conrad)
Pholadomya sp.
Pitar uvasanus (Conrad)
Schedocardia cf. *S. brewerii* (Gabb)
Amaurellina aff. *A. cortezia* Gardner and Bowles
Amaurellina aff. *A. moragai* Stewart
Ectinochilus canalifer supraplicatus (Gabb)
Ficopsis hornii (Gabb)
Ficus mamillatus (Gabb)
Galeodea susanae Schenck
Murex? whitneyi (Gabb)
Olequahia cf. *O. hornii* (Gabb)
Seraphs erratica (Cooper)
Surculites sinuatus (Gabb)
Turritella scrippsensis Hanna
Voluta martini Dickerson

Kelley recognized the mixture of "Tejon" and "Domengine" elements in this assemblage and referred it to the "Transition Stage." Kleinpell and Weaver (1963: 106) also referred this assemblage to the "Transition Stage" and considered *Turritella scrippsensis* to be restricted to this "Stage." In the Pine Mountain section, however, *T. scrippsensis* is associated with taxa characteristic of the "Domengine Stage." Thus, its range extends down into that "Stage."

Ectinochilus canalifer FAUNA

The *Ectinochilus canalifer* fauna is referable to the "Tejon Stage." It shares many taxa in common with the fauna of the type Tejon Formation, including the two taxa, *Turritella uvasana* s.s. and *T. uvasana sargeanti,* that have been regarded as index fossils for the "Tejon Stage" (Clark and Vokes, 1936:853, 867; Weaver et al., 1944, pl. 1; Kleinpell and Weaver, 1963:106, 107). Many other taxa that are restricted to the *Ectinochilus canalifer* fauna in the Pine Mountain section, including *Turritella schencki* s.s., *Siphonalia sopenahensis, Perse sinuata, Olequahia hornii, Molopophorus tejonensis, Hemipleurotoma pulchra, Gemmula abacta, Conus hornii* s.s., *Exilia fausta, Acteon quercus, Architectonica hornii, Venericardia hornii* s.s., *Tellina tehachapi, T. lebecki, T. castacana, Spisula bisculpturata, Pitar californianus, Callista hornii* s.s., *Crassatella uvasana* s.s., and *Glycymeris viticola,* have been reported elsewhere on the Pacific Coast only from strata assigned to the "Tejon Stage" (Weaver, 1912; 1942; Dickerson, 1915; Weaver and Palmer, 1922; Anderson and Hanna, 1925; Hanna, 1927; Turner, 1938; Clark, 1938; Merriam, 1941; Kleinpell and Weaver, 1963).

Several taxa that are restricted to the *Ectinochilus canalifer* fauna in the Pine Mountain section, including *Neverita secta, Ficopsis remondii* s.s., *Ectinochilus canalifer, Thracia dilleri, Pitar dickersoni, Callista andersoni,* and *Corbula hornii,* have been recorded from "Tejon" and younger strata (viz., the *Turritella schencki delaguerrae* and *Turritella variata lorenzana* Zones of Kleinpell and Weaver, 1963) elsewhere on the Pacific Coast but have not been recorded from strata older than the "Tejon Stage." Other taxa that are restricted to or have their highest occurrence within this fauna in the Pine Mountain section, including *Turritella uvasana neopleura, T. buwaldana, Nekewis io, Turricula cohni, Terebra californica, Natica uvasana, Euspira nuciformis, Strepsidura ficus, Pachycrommium? clarki, Sinum obliquum, Pseudoliva inornata, Ranellina pilsbryi, Laevityphis antiquus, Cylichnina tantilla, Hexaplex? whitneyi, Ficus mamillata, Pachecoa hornii, Nemocardium linteum, Pitar uvasanus, Cyclinella elevata, Macoma viticola,* and *Nuculana gabbi,* have been recorded elsewhere from "Tejon" or older strata but have not been recorded from strata younger than the "Tejon Stage." The joint occurrence of these taxa is additional evidence for the "Tejon" age of the *Ectinochilus canalifer* fauna.

Four species in the *Ectinochilus canalifer* fauna, including *Loxotrema turritum, Nerita triangulata, Potamides carbonicola,* and *"Trichotropis" lajollaensis,* previously have not been reported from strata younger than the "Domengine Stage." Their occurrence in this fauna, stratigraphically above faunas referable to the "Domengine" and "Transition" "Stages," is evidence that their ranges extend into the "Tejon Stage."

Stage Nomenclature and Stage Boundaries

The faunal data from the Pine Mountain section confirm the existence and sequence of the traditional Eocene "Stages" of the Pacific Coast and permit more confident recognition of these units than has heretofore been possible. Each "Stage" can be recognized by the restricted occurrence of some taxa and the concurrent ranges of others. The data also indicate, however, that the "Domengine Stage" and the upper part of the "Capay Stage" (i.e., the *Galeodea susanae* Zone) are equivalent and that some taxa formerly regarded as index fossils are not restricted to the units that they are supposed to characterize. Further revision and refinement of the Eocene "Stages" are necessary before they can be accepted as formal chronostratigraphic units.

The current nomenclature of these "Stages" is unsatisfactory. Three of the "Stages" ("Capay," "Domengine," and "Tejon") have been named after lithostratigraphic units (formations); the name "Transition Stage" was never proposed as a formal Stage name (see Clark and Vokes, 1936:863). In order to avoid confusion between chronostratigraphic and lithostratigraphic units, the American Commission on Stratigraphic Nomenclature (1970:15, Article 32-c) recommends that Stages be given geographic names that have not been previously used in stratigraphic classification. Therefore, at such a time as the "Capay," "Domengine," and "Tejon" "Stages" are formally defined, they should be renamed. The name "Transition" should also be abandoned and replaced by a geographic name.

The boundaries of the Eocene "Stages" are also poorly defined. Previous authors (Clark, 1926; 1935; 1943; Clark and Vokes, 1936; Merriam and Turner, 1937; Vokes, 1939) considered the boundaries of these "Stages" to be coincident with formation boundaries, even though the characteristic fauna of a given "Stage" may be confined to only a portion of the corresponding formation. Thus, for example, the "Capay Stage" was proposed by Clark (1935:1036n) for strata contemporaneous with those of the type Capay Formation. As discussed earlier in this paper, however, the characteristic fauna of the "Capay Stage" has been recorded only from the lower part of the type Capay Formation. The upper part of the formation has not yielded a molluscan fauna and may be younger than "Capay" age. The Rose Canyon Shale near San Diego, California, provides another example. This unit, designated by Clark (1943:188) as the type section of the "Transition Stage" or *"Rimella supraplicata* Zone," actually contains mollusks characteristic of both the "Domengine" and "Transition" "Stages." Thus, a new type section must be designated for the "Transition Stage." Pending the formal definition of this unit, the strata containing the *Ectinochilus supraplicatus* fauna in the Pine Mountain section are designated as a reference section.

Precise biostratigraphic limits are lacking for the Eocene "Stages." The "Zones" assigned to these units by previous authors represent little more than local assemblage-zones. They overlap in faunal composition and their boundaries are indefinite. Few of the "index fossils" recognized by previous authors are restricted to the zones they are supposed to characterize.

I suggest that future biostratigraphic studies of Eocene mollusks on the Pacific

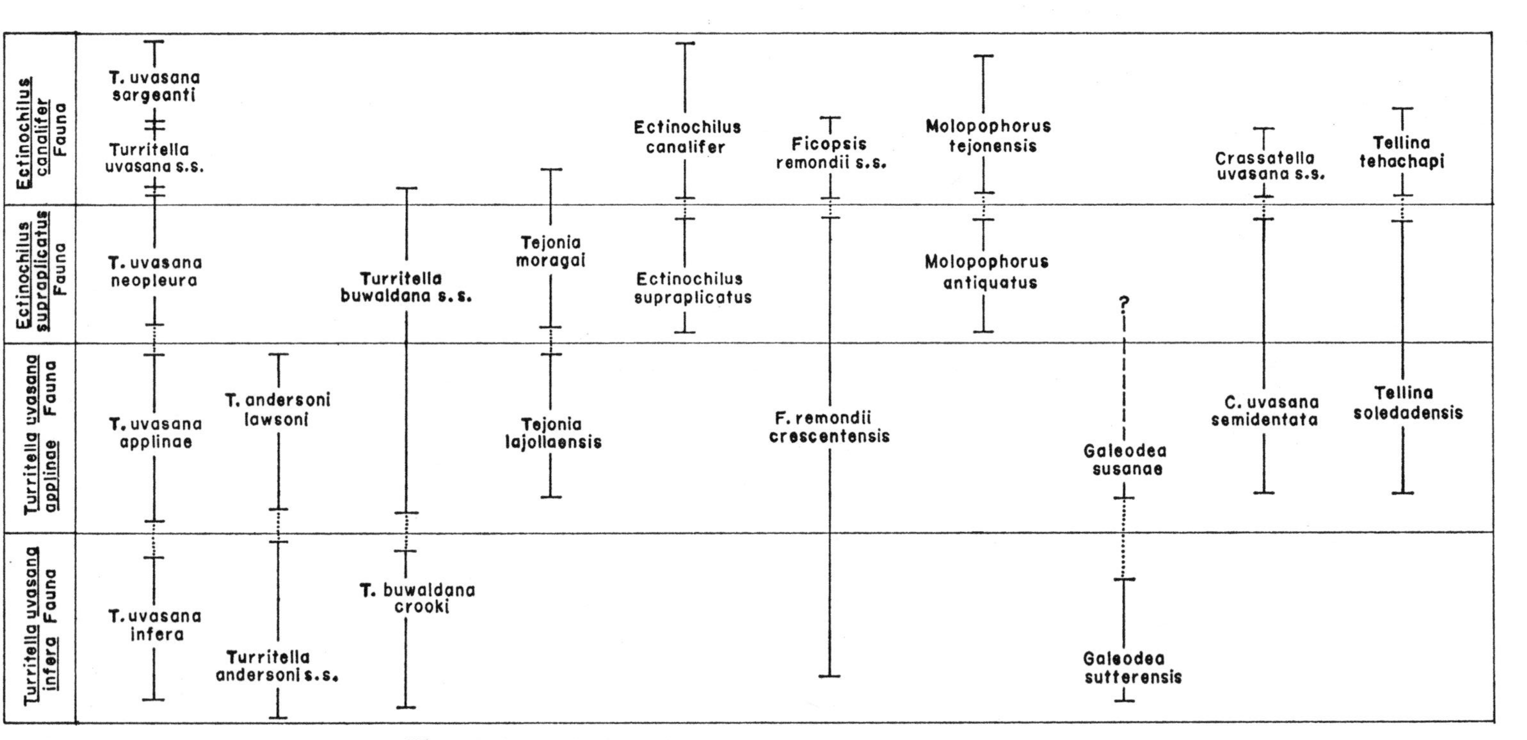

Fig. 7. Inferred phylogenetic sequences in the Pine Mountain section.

Coast be directed toward establishing a zonation based upon evolutionary events and that the limits of these zones be used to define formal Stage boundaries. Apparent evolutionary sequences have been recognized within several molluscan taxa in the Pine Mountain section (fig. 7). Most of these sequences have also been suggested by previous authors (Clark and Vokes, 1936:863, 864; Turner, 1938:86; Vokes, 1939:32, 37, 38; Merriam, 1941:43, 52). Prior to this study, however, adequate stratigraphic confirmation has been lacking. Of particular significance for correlation are the apparent phylogenetic sequences within *Turritella uvasana* and *Turritella andersoni.* These two species are among the most abundant and widely distributed elements in the Eocene molluscan faunas of the Pacific Coast. Careful study of their phylogeny, therefore, may provide a basis for the definition of precise, chronologically significant, and widely recognizable Stage boundaries.

DEPOSITIONAL ENVIRONMENTS OF THE PINE MOUNTAIN EOCENE FORMATIONS

The Eocene formations in the Pine Mountain area were deposited in a variety of sedimentary environments, including continental, deltaic, and marine. These formations are inferred to have been deposited within and adjacent to a large marine delta complex along the northeastern margin of an Eocene marine embayment in the western Transverse Ranges.

Juncal Formation

Lithologic features and fossils indicate that the Juncal Formation in the Pine Mountain area was deposited in continental, transitional, and marine environments. The mudstone facies is entirely of marine origin. The conglomerate facies, on the other hand, appears to be mainly of fluvial (continental) origin, although portions of it may also have been deposited directly in a shallow marine or deltaic environment. The sandstone and siltstone facies contain evidence of deposition in both fluvial and marine environments and are interpreted to represent the deposits of a large marine delta complex.

Mudstone facies.—The presence of abundant mollusks, including such genera as *Turritella, Glycymeris, Calyptraea, Sinum, Corbula, Gari, Pitar, Tellina, Cerithium,* and *Arca,* in the lower 500 ft of the mudstone facies of the Juncal Formation indicates that these strata were deposited in a nearshore shallow marine environment. Living species of these genera commonly occur in inner sublittoral environments along the Pacific Coast of North America (Keen, 1958; 1963:99–109; Parker, 1964). Above the mollusk-bearing strata, the only fossils noted in the remainder of the mudstone facies were occasional foraminifera. Since these were not examined during this study, the depth at which the bulk of the mudstone facies was deposited is unknown. The very fine-grained texture and uniform lithology of the mudstone and the general absence of sedimentary structures indicative of shallow-water deposition suggest that much of this facies was deposited below wave base in a calm outer sublittoral or bathyal environment.

Conglomerate facies.—This facies appears to be largely of fluvial origin, although portions of it may have been deposited in a shallow marine environment. Lithologic features of the conglomerate which are indicative of a fluvial origin (see Barrell, 1925; Pettijohn, 1957; Allen, 1965) include (1) great thickness, (2) coarse texture, (3) generally poor sorting, (4) polymictic composition (including plutonic, volcanic, metamorphic, and sedimentary clasts), (5) clast imbrication, and (6) rapid wedging out into finer-grained deposits. These features, according to Pettijohn (1957:257–260), are characteristic of "thick, wedge-shaped basin-margin accumulations of gravel that were shed from sharply elevated highlands." A shallow marine origin for a part of the conglomerate facies, on the other hand, is suggested by its intricately interbedded and intertongued relationship with the deltaic and marine sandstone, siltstone, and mudstone facies of the Juncal Formation. Kiessling (1958) reported fragments of oysters and other unidentifiable pelecypods from one of the conglomerate lenses near Grade Valley and oyster fragments were noted in the conglomerate facies at several localities during this study. It is possible, however, that these fossils have been reworked from the other facies of the formation. On the basis of its predominantly fluvial character and its intimate association with marine and deltaic deposits, the conglomerate facies is interpreted to represent stream channel deposits that were laid down on an alluvial or deltaic coastal plain adjacent to the Eocene marine embayment in the western Transverse Ranges. Locally, some of the conglomerate may have been deposited directly in a shallow marine or littoral environment.

The coarse texture and the composition of the conglomerate indicate that it was derived from a nearby mountainous source area composed predominantly of granitic and high-grade metamorphic rocks. It was probably derived in large part from erosion of the crystalline rocks of the Basement Complex east and northeast of the mapped area. An eastern or northeastern source area for the conglomerate is indicated by its southward and westward lensing into finer-grained strata and by clast imbrication measurements by Jestes (1963) which indicate a southwestward direction of transport for a portion of the conglomerate facies. Most of the plutonic and metamorphic rock types in the conglomerate occur in the Basement Complex east of the mapped area (Schlee, 1952; Carman, 1964). The source of the volcanic clasts, however, is unknown. Volcanic rocks of pre-Eocene age have not been reported from the surrounding region. These clasts, and perhaps also many of the durable quartzite clasts, may have been recycled from older, pre-Eocene, conglomerates.

Sandstone facies.—This facies, like the conglomerate facies, displays lithologic features typically associated with fluvial deposits (see Doeglas, 1962; Allen, 1965; Coleman and Gagliano, 1965), including large-scale *planar-* and *trough-*type cross-stratification; very thick, markedly lenticular, bedding; poor sorting; coarse texture, including gravel lenses; scour-and-fill structure; and intraformational conglomerate. Much of this facies resembles point bar and channel fill deposits of meandering streams. Locally, however, the sandstone facies contains fossil mollusks, including such genera as *Turritella, Architectonica, Sinum, Olivella, Natica, Ostrea, Potamides, Spisula, Barbatia, Isognomon, Brachidontes, Gari, Corbula, Tellina, Macoma, Glycymeris, and Pitar* (table 1, UCR localities 4752, 4673, 4674, 4695, 4698, 4717, 4718, 4722), which are indicative of deposition in a shallow marine (inner sublittoral) environment (Keen, 1958; 1963; Abbott, 1958; Warmke and Abbott, 1962; Parker, 1964). Living species of *Isognomon* and *Brachidontes* on the Pacific Coast of North America and in the Caribbean commonly occur intertidally and in very shallow water attached to rocks and mangrove roots (Keen, 1958:48, 62; Warmke and Abbott, 1962:162, 165). *Barbatia* also commonly occurs under rocks in the intertidal zone (Keen, 1958:26–30; 1963: 105; Warmke and Abbott, 1962:158). Oysters attain their optimum development in very shallow water and some living species (e.g., *Crassostrea virginica* and *Ostrea edulis*) can tolerate brackish-water conditions (Stenzel, 1971:1038–1039). According to Stenzel (ibid., p. 1041), only coastal brackish-water oyster species form reefs. The occurrence of oyster banks or "reefs" in the green mudstone beds (e.g., UCR localities 4695, 4698) in the upper part of the sandstone facies, therefore, suggests that these strata may have been deposited in brackish-water bays or lagoons similar to those along the present-day Texas coast (Ladd, 1951; Hedgpeth, 1953; Parker, 1959). The occurrence of abundant specimens of the gastropod *Potamides* in association with *Ostrea* at locality 4698 also suggests brackish-water conditions. Living members of the family Potamididae typically occur on mud flats and in mangrove swamps in brackish-water bays and lagoons (Keen, 1958:306–307; Warmke and Abbott, 1962:71). The local occurrence of callianassid burrows in the sandstone facies also suggests a shallow marine or littoral origin for portions of this unit. These burrows are particularly common in the fine- to medium-grained, moderately sorted, sandstone beds interbedded with the oyster-bearing mudstones in the upper part of the sandstone facies. Living species of *Callianassa* are common inhabitants of littoral and inner sublittoral environments along the Pacific, Gulf, and Atlantic coasts of the United States (McGinnitie and McGinnitie, 1949; Rickets and Calvin, 1960; Weimer and Hoyt, 1964; Warme, 1968).

On the basis of its mixed fluvial-marine character, the sandstone facies is interpreted to be of deltaic origin. Most of this facies probably represents distributary stream channel deposits laid down on a subaerial deltaic plain. The sandstone beds containing the marine mollusks and the callianassid burrows, on the other hand, may represent beach and bar deposits formed at the seaward margin of the delta complex. The oyster-bearing mudstones were probably deposited in brackish-water bays or lagoons.

Siltstone facies.—On the basis of its intricately interbedded and intertongued relationship with the sandstone facies, the siltstone facies is interpreted to represent the interdistributary overbank or topstratum (Allen, 1965) deposits of the delta complex. The fine-grained texture and poor sorting of the siltstone and the local occurrence of abundant fossil plant debris are consistent with this interpretation. Most of the siltstone lacks fossils other than plant debris and was probably deposited on the subaerial deltaic plain. The local occurrence of abundant mollusks, including such genera as *Pitar, Glycymeris, Solariella, Architectonica, Ostrea, Turritella, Natica, Cerithium, Corbula, Sinum, Calyptraea, Conus, Homalopoma, Terebra, Pyramidella, Tellina,* and *Brachidontes,* indicates, however, that part of the siltstone facies was deposited in a shallow marine (inner sublittoral) environment (Keen, 1958; 1963; Parker, 1964). The strata containing these fossils may have been deposited in interdistributary bays or lagoons or in the prodelta environment (Coleman and Gagliano, 1965:143) at the seaward margin of the deltaic complex.

Matilija Sandstone

The presence of abundant mollusks, including such genera as *Turritella, Architectonica, Sinum, Polinices, Neverita, Natica, Calyptraea, Crepidula, Conus, Olivella. Ostrea, Crassatella, Glycymeris, Corbula, Tellina, Macoma, Terebra, Pitar, Nerita,* and *Brachidontes,* in the Matilija Sandstone indicates that this unit was deposited in a nearshore, inner-sublittoral, marine environment. In the Gulf of California, recent species of most of these genera were recorded by Parker (1964: 344–351, table 1) as living together on sandy to muddy substrates at depths ranging from intertidal to about 100 ft. *Nerita* and *Brachidontes* were recorded by Parker (ibid., pp. 344–346, table 1) only from intertidal rocky shore habitats. The local occurrence of oyster beds (UCR locality 4724) and the brackish-water gastropods *Potamides* and *Loxotrema* (UCR localities 4719, 4720, 4721, 4722) suggest that portions of the Matilija Sandstone may have been deposited in brackish-water bays or lagoons (Keen, 1958:306; Stenzel, 1971:1041).

The Matilija Sandstone intertongues laterally with the sandstone facies of the Juncal Formation, indicating that it is closely related in origin to the Juncal delta complex. On the basis of this relationship and its shallow marine fossils, the Matilija Sandstone is interpreted to represent sediments deposited in delta front and prodelta environments (Coleman and Gagliano, 1965:143) at the seaward margin of the delta complex. Coleman and Gagliano (ibid., pp. 143–145) recognized several subenvironments within the delta front environment, including distal bar, distributary mouth bar, channel, and subaqueous levee. Most or all of these subenvironments may be represented in the Matilija Sandstone, but more detailed studies are required for their recognition.

Cozy Dell Shale

The occurrence of the molluscan genera *Turritella* and *Euspira* (UCR localities 4663, 4730) in the Cozy Dell Shale indicates that this unit was deposited in a marine environment. The fine-grained lithology of the Cozy Dell and its intertonguing relationship with coarser-grained inner-sublittoral deposits of the Matilija Sandstone suggest that it was deposited below wave base in a calm offshore, outer-sublittoral or bathyal environment.

Coldwater Sandstone

The coarse texture of the Coldwater Sandstone and the local occurrence of shallow marine and brackish-water mollusks such as *Venericardia, Ostrea,* and *Corbicula* (table 1, UCR localities 4730 and 4663) suggest that this unit was deposited in very nearshore marine and marginal marine environments. The thick-bedded, cross-stratified, well-sorted, medium- to coarse-grained sandstone that makes up the bulk of the formation in the Pine Mountain area probably represents beach and bar deposits. The few thick beds of very coarse-grained, conglomeratic sandstone in the unit closely resemble the sandstone facies of the Juncal Formation and may represent deltaic distributary channel deposits. These beds contain sedimentary structures typically associated with channel-fill deposits, including large-scale *planar* and *trough* cross-stratification, scour-and-fill and intraformational conglomerate. The occurrence of *Ostrea* and *Corbicula* in the brownish green silty mudstone beds in the upper part of the Coldwater Sandstone near Pine Mountain Lodge (see fig. 5, UCR locality 4663) suggests that these strata may have been deposited in a brackish-water bay or lagoon. According to Keen (1958:88), members of the Family Corbiculidae typically occur in brackish to freshwater habitats.

Eocene Climates

The Eocene molluscan faunas in the Pine Mountain area are indicative of shallow-water marine climates warmer than exist at this latitude today. Recent species of many taxa in these faunas, including *Isognomon, Nerita (Theliostyla), Xenophora, Callista (Macrocallista), Pitar (Lamelliconcha), Crassatella, Architectonica, Ficus, Cerithium, Brachidontes, Arca, Cyclinella, Hexaplex, Sassia, Pseudoliva, Terebra, Barbatia, Conus,* and *Pteria,* are largely or entirely confined to tropical or subtropical faunal provinces (Abbott, 1954; Keen, 1958; 1963; Warmke and Abbott, 1962; Parker, 1964). Several of the extinct generic or subgeneric taxa in these faunas, including *Laevityphis, Ficopsis, Volutocristata, Campanilopa, Eocernina, Eotibia, Gomphopages, Coalingodea, Chedevillea,* and *Ectinochilus,* have modern analogues that are also tropical or subtropical in distribution.

SYSTEMATIC CATALOG

The identification of the species listed in this catalog is based mainly on published figures and descriptions, although many of the identifications have been checked by comparison with topotypes on deposit in the geology departments at the Riverside and Los Angeles campuses of the University of California. Comparison has also been made with Gabb's (1864; 1869) types at the Academy of Natural Sciences of Philadelphia and with some of Turner's (1938) and Vokes's (1939) types in the University of California Museum of Paleontology, Berkeley, California. Hypotypes of the species identified during this study are on deposit in the Department of Geological Sciences, University of California, Riverside.

No attempt has been made to include a complete synonymy for the species listed herein. Rather, with few exceptions, only those works that include figures and/or descriptions are listed. More complete synonymies and discussions of these species can be found in the following important reference works on the West Coast Eocene: Gabb (1864; 1869), Dickerson (1913; 1914; 1915; 1916), Weaver and Palmer (1922), Anderson and Hanna (1925), Stewart (1926; 1930), Hanna (1927), Clark and Woodford (1927), Turner (1938), Clark (1938), Vokes (1939), Merriam (1941), and Weaver (1942).

The systematic arrangement of the generic and higher taxa generally follows that of Cox et al. (1969) for the pelecypods and Wenz (1938–1944) for the gastropods.

The provincial ranges of the species listed herein have been compiled from data in the Pine Mountain section and from published records for other Lower Tertiary sections on the West Coast, including the Lodo and Domengine Formations north of Coalinga, California (Vokes, 1939); the type Tejon Formation (Dickerson, 1915; Anderson and Hanna, 1925; Marks, 1941; 1943); the La Jolla and Poway Groups near San Diego, California (Hanna, 1927; Kennedy and Moore, 1971); the Matilija, Cozy Dell, Sacate, Coldwater, and Gaviota Formations west of Santa Barbara, California (Kelley, 1943; Kleinpell and Weaver, 1963); the Avenal Sandstone on Reef Ridge, south of Coalinga, California (Stewart, 1946); the type Meganos Formation, north of Mount Diablo (Clark and Woodford, 1927); the Muir Sandstone near Martinez, California (Weaver, 1953); the Markley Sandstone north of Vacaville, California (Clark, 1938); the type Capay Formation in Capay Valley, California (Merriam and Turner, 1937); the Marysville Claystone Member of the Meganos Formation at Marysville Buttes in the Sacramento Valley (Dickerson, 1913; 1916; Stewart, 1949); the Umpqua, Tyee, Coaledo, and Spencer Formations in western Oregon (Turner, 1938); the Eugene Formation in western Oregon (Hickman, 1969); and the Cowlitz Formation in western Washington (Weaver, 1942).

Phylum MOLLUSCA
Class BIVALVIA
Superfamily NUCULACEA
Family NUCULIDAE
Genus *Acila* Adams and Adams, 1858

Type species: (by subsequent designation, Stoliczka, 1871) *Nucula divaricata* Hinds

Subgenus *Truncacila* Schenck, *in* Grant and Gale, 1931

Type species: (by original designation) *Nucula castrensis* Hinds.

Acila (*Truncacila*) *decisa* (Conrad)
(Pl. 1, fig. 1)

Nucula decisa Conrad, 1857, pl. 3, fig. 19.
Nucula (*Acila*) *stillwaterensis* Weaver and Palmer, 1922:6, pl. 8, fig. 8.
Acila gabbiana Dickerson, 1916:481, pl. 36, fig. 1. Anderson and Hanna, 1925:176, pl. 9, fig. 12.
Acila lajollaensis Hanna, 1927:270, pl. 25, figs. 1, 3, 5, 7, 8, 12, 15.
Acila (*Truncacila*) *decisa* (Conrad). Schenck, 1936:53–56, pl. 3, figs. 1–9, 11–15; pl. 4, figs. 1–2; text fig. 7 (22–27). Turner, 1938:41, 42, pl. 5, figs. 2, 3. Vokes, 1939:41, pl. 1, figs. 7, 8.

Hypotypes: UCR 4667/7, locality 4667; UCR 4675/231, locality 4675.

Local occurrence: Turritella uvasana infera and *T. uvasana applinae* faunas, Juncal Formation.

Provincial Range: Late Paleocene ("Meganos Stage") to late Eocene (*Turritella schencki delaguerrae* Zone of Kleinpell and Weaver, 1963).

Superfamily NUCULANACEA
Family NUCULANIDAE
Genus *Nuculana* Link, 1807

Type species: (by original designation) *Arca rostrata* Chemnitz, 1774 (= *Arca pernula* Müller, 1771).

Subgenus *Nuculana* s.s.

Nuculana cf. *N. washingtonensis* (Weaver)

Hypotype: UCR 4727/81, locality 4727, *Ectinochilus canalifer* fauna, Matilija Sandstone.

Several small, poorly preserved specimens of a *Nuculana* s.s. closely resembling Weaver's (1916:34,35, pl. 3, figs. 25, 26) species were collected from locality 4727.

Subgenus *Saccella* Woodring, 1925

Type species: (by original designation) *Arca fragilis* Chemnitz.

Nuculana (*Saccella*) *gabbii* (Gabb)

(Pl. 1, fig. 3)

Leda protexta? Gabb, 1864:199 (in part), pl. 26, fig. 185.
Leda vogdesi Anderson and Hanna, 1925:177, pl. 2, figs. 8, 9.
Saccella gabbii (Gabb). Stewart, 1930:55, 56, pl. 7, fig. 3; pl. 10, fig. 4.
Nuculana (*Saccella*) *gabbii* (Gabb). Vokes, 1939: 41, 42.

Hypotype: UCR 4743/301, locality 4743.

Local occurrence: Ectinochilus canalifer fauna, Matilija Sandstone.

Provincial Range: Late Paleocene ("Meganos Stage") to late Eocene (*Turritella schencki delaguerrae* Zone of Kleinpell and Weaver, 1963).

Nuculana (*Saccella*) *hondana* Vokes

(Pl. 1, fig. 4)

Nuculana (*Saccella*) *hondana* Vokes, 1939: 42, 43, pl. 1, figs. 9, 10.

Hypotype: UCR 4662/12, locality 4662.

Local occurrence: Turritella uvasana infera fauna, Juncal Formation.

Provincial Range: Early Eocene ("Capay Stage").

Slightly deformed specimens referable to this small species were collected from several localities in the lower part of the mudstone facies of the Juncal Formation. All show the bluntly angulate rostrum, the smooth shallow groove extending from the beaks to the posterior point of the rostral margin, and the sculpture of narrow, widely and regularly spaced concentric lines that are characteristic of *N. hondana*.

Nuculana (*Saccella*) *parkei* (Anderson and Hanna) s.s.

Leda parkei Anderson and Hanna, 1925:179, 180, pl. 2, figs. 10, 11.

Hypotype: UCR 4703/10, locality 4703.

Local occurrence: Turritella uvasana infera and *T. uvasana applinae* faunas, Juncal Formation.

Provincial Range: Early Eocene ("Capay Stage") to late Eocene ("Tejon Stage").

Genus *Ledina* Dall, 1898

Type species: (by original designation) *Leda eborea* Conrad, 1860 (= *Leda smirna* Dall, 1898).

Ledina fresnoensis (Dickerson)

(Pl. 1, fig. 2)

Leda fresnoensis Dickerson, 1916:483, 484, pl. 36, figs. 2*a*, 2*b*.
Nuculana fresnoensis (Dickerson). Merriam and Turner, 1937:94, 96, 99. Turner, 1938:42, pl. 5, fig. 6.

Calorhadia (*Litorhadia*) *fresnoensis* (Dickerson). Vokes, 1939:43, 44, pl. 1, fig. 5.
Hypotype: UCR 4662/101, locality 4662.
Local occurrence: Turritella uvasana infera fauna, Juncal Formation.
Provincial Range: Early Eocene ("Capay Stage") to middle Eocene ("Domengine Stage").

This species is referred to *Ledina* on the basis of its nearly smooth shell, equilateral shape, and evenly rounded anterior and posterior margins.

Genus *Portlandia* Mörch, 1857

Type species: (by subsequent designation, International Comm. Zool. Nomen., Opinion 769, 1966) *Nucula arctica* Gray, 1824.

Subgenus *Portlandia* s.s.

Portlandia (*Portlandia*) *rosa* (Hanna)

Leda rosa Hanna, 1927:271, pl. 25, figs. 4, 6, 9, 16.
Portlandia (*Portlandella*) *rosa* (Hanna). Stewart, 1930:61, 62.
Nuculana rosa (Hanna). Keen and Bentson, 1944:57.
Hypotype: UCR 4699/12, locality 4699.
Local occurrence: Turritella uvasana applinae fauna, Juncal Formation.
Provincial Range: Middle Eocene ("Domengine Stage").

A single specimen of this species was found at locality 4699. Stewart (1930:61) designated *Leda rosa* Hanna as the type species of a new subgenus *Portlandella.* Cox et al. (1969, part N, 1:239), however, regard *Portlandella* to be a synonym of *Portlandia* s.s.

Superfamily ARCACEA

Family ARCIDAE

Subfamily ARCINAE

Genus *Arca* Linné, 1758

Type species: (by subsequent designation, Schmidt, 1818) *Arca noae* Linné.

Subgenus *Arca* s.s.

Arca (Arca) n. sp.?

(Pl. 1, fig. 8)

Hypotype: UCR 4667/131, locality 4667, *Turritella uvasana infera* fauna, Juncal Formation.

Several small, somewhat poorly preserved specimens of an *Arca* s.s. collected from locality 4667 probably represent a new species. They resemble *Arca hawleyi* Reinhart (1943:21, pl. 2, figs. 20, 22) from the late Eocene of California, but are only about one-sixth as large and differ slightly in outline. Reinhart (1943: 22) describes the shell of *A. hawleyi* as narrowing posteriorly, whereas the shells from locality 4667 broaden posteriorly.

Genus *Barbatia* Gray, 1842

Type species: (by subsequent designation, Gray, 1857) *Arca barbata* Linné.

Subgenus *Barbatia* s.s.

Barbatia (*Barbatia*) *morsei* Gabb

Barbatia morsei Gabb, 1864:216, pl. 32, fig. 286. Arnold, 1909:108, pl. 3, fig. 8. Hanna, 1927:272, 273, pl. 25, figs. 2, 10, 11, 13, 14. Stewart, 1930:87, 88, pl. 8, fig. 7.
Barbatia (*Obliquarca*) *morsei* Gabb. Vokes, 1939:49, pl. 1, figs. 25, 26, 28, 29. Reinhart, 1943:30, 31, pl. 1, fig. 4.

Hypotype: UCR 4752/7, locality 4752.

Local occurrence: Turritella uvasana applinae fauna, Juncal Formation.

Provincial Range: Middle Eocene ("Domengine Stage").

Cox et al. (1969, part N, 1:252) regard *Obliquarca* Sacco, 1898 (type species *Arca modioliformis* Deshayes, 1831) to be a synonym of *Barbatia* s.s.

Family PARALLELODONTIDAE

Subfamily GRAMMATODONTINAE

Genus *Porterius* Clark, 1925

Type species: (by original designation) *Barbatia andersoni* Van Winkle.

Porterius woodfordi (Hanna)

Barbatia woodfordi Hanna, 1927:273, pl. 27, figs. 1, 6, 8, 10.

Porterius woodfordi (Hanna). Reinhart, 1937:176, 177. Vokes, 1939:45, pl. 1, fig. 13.

Hypotype: UCR 4658/12, locality 4658.

Local occurrence: Turritella uvasana infera fauna, Juncal Formation.

Provincial Range: Early Eocene ("Capay Stage") to middle Eocene ("Domengine Stage").

A single specimen representing this species was collected from locality 4658.

Family NOETIIDAE

Subfamily TRINACRIINAE

Genus *Pachecoa* Harris, 1919

Type species: (by original designation) *Trinacria (P.) cainei* Harris.

Subgenus *Pachecoa* s.s.

Pachecoa (Pachecoa) hornii (Gabb) s.s.

Arca hornii Gabb, 1864:194, pl. 30, fig. 263. Dickerson, 1915:78, pl. 1, fig. 4; 1916:504, pl. 36, fig. 4. Anderson and Hanna, 1925:180, pl. 2, figs. 1, 2.

Halonanus hornii (Gabb). Stewart, 1930:79, 80, pl. 10, fig. 6.

Trigonodesma hornii (Gabb). Reinhart, 1935:53. Vokes, 1939: 49, pl. 1, figs. 24, 27.

Halonanus hornii subsp. *hornii* (Gabb). Reinhart, 1943:79, 80, pl. 2, figs. 7–9.

Hypotypes: UCR 4706/16, locality 4706; UCR 4721/2, locality 4721.

Local occurrence: Ectinochilus supraplicatus and *E. canalifer* faunas, Juncal and Matilija formations.

Provincial Range: Middle Eocene ("Domengine Stage") to late Eocene ("Tejon Stage").

Cox et al. (1969, part N, 1:264) regard *Halonanus* Stewart, 1930, to be a synonym of *Pachecoa* Harris s.s.

Superfamily LIMOPSACEA

Family LIMOPSIDAE

Genus *Limopsis* Sassi, 1827

Type species: (by original designation) *Arca aurita* Brocchi.

Subgenus *Limopsis* s.s.

Limopsis (Limopsis) marysvillensis (Dickerson)

(Pl. 1, fig. 7)

Glycimeris[2] *marysvillensis* Dickerson, 1913:290, pl. 14, figs. 1*a*, 1*b*; 1916, pl. 40, figs. 9*a*, 9*b*.

[2] Error for *Glycymeris*.

Limopsis marysvillensis (Dickerson). Merriam and Turner, 1937, table 2.

Hypotype: UCR 4656/101, locality 4656.

Local occurrence: Turritella uvasana infera fauna, Juncal Formation.

Provincial Range: Early Eocene ("Capay Stage").

Family GLYCYMERIDIDAE

Subfamily GLYCYMERIDINAE

Genus *Glycymeris* da Costa, 1778

Type species: (by tautonymy) *Arca orbicularis* da Costa, 1778 (= *Arca glycymeris* Linné, 1758).

Subgenus *Glycymeris* s.s.

Glycymeris (*Glycymeris*) aff. *G.* (*G.*) *fresnoensis* Dickerson

(Pl. 1, fig. 14)

Hypotype: UCR 4661/91, locality 4661, *Turritella uvasana infera* fauna, Juncal Formation.

Two specimens from locality 4661 closely resemble *G. fresnoensis* in having strong, bifurcating, radial costae, but are more quadrate in outline, have a broader hinge line, and have less prominent umbones. Whether these differences are of taxonomic significance cannot be determined until more is known about the range of variation of *G. fresnoensis*.

Glycymeris (Glycymeris) n. sp.? aff. *G.* (*G.*) *perrini* Dickerson

(Pl. 1, fig. 5)

Hypotype: UCR 4697/301, locality 4697, *Ectinochilus supraplicatus* fauna, Juncal Formation.

Several well-preserved specimens of *Glycymeris* collected from locality 4697 resemble *G. perrini* Dickerson (1916:482, pl. 36, figs. 6*a*–6*c*) (see also Vokes, 1939:47, pl. 1, fig. 23) in shape and ornamentation, but attain a larger size than appears to be typical for that species and have less prominent umbones. They may represent a new species.

Glycymeris (*Glycymeris*) *rosecanyonensis* Hanna

Glycymeris rosecanyonensis Hanna, 1927:273, 274, pl. 27, figs. 4, 5, 9, 11.

Hypotypes: UCR 4680/221, locality 4680; UCR 4681/101, locality 4681.

Local occurrence: *Turritella uvasana applinae* and *Ectinochilus supraplicatus* faunas, Juncal Formation.

Provincial Range: Middle Eocene ("Domengine" and "Transition" "Stages").

Glycymeris (*Glycymeris*) *viticola* Anderson and Hanna

(Pl. 1, fig. 6)

Glycymeris viticola Anderson and Hanna, 1925:182, 183, pl. 1, fig. 5; pl. 3, fig. 1.

Hypotype: UCR 4723/141, locality 4723.

Local occurrence: *Ectinochilus canalifer* fauna, Matilija Sandstone.

Provincial Range: Late Eocene ("Tejon Stage").

Subgenus *Glycymerita* Finlay and Marwick, 1937

Type species: (by original designation) *Glycymeris concava* Marshall.

Glycymeris (*Glycymerita*) *sagittata* (Gabb)

Axinaea (*Limopsis?*) *sagittata* Gabb, 1864:197, 198, pl. 31, figs. 267, 267*a*.

Glycymeris sagittata (Gabb). Anderson and Hanna, 1925:181, 182, pl. 1, fig. 6. Stewart, 1930:71–73, pl. 12, fig. 10. Vokes, 1939:45, 46, pl. 1, figs. 18–20. Kleinpell and Weaver, 1963:196, 197, pl. 28, fig. 10; pl. 29, figs. 1, 2.

Glycimeris sagittatus (Gabb). Turner, 1938:43, 44, pl. 6, figs. 1–3.

Hypotypes: UCR 4680/12, locality 4680; UCR 4706/27, locality 4706; UCR 4721/18 and 4744/8, localities 4721 and 4744, respectively.

Local occurrence: ***Turritella uvasana applinae, Ectinochilus supraplicatus,*** **and *E. canalifer* faunas, Juncal and Matilija formations.**

Provincial Range: Early Eocene ("Capay Stage") to late Eocene (***Turritella variata lorenzana*** Zone of Kleinpell and Weaver, 1963).

Superfamily MYTILACEA
Family Mytilidae
Subfamily Mytilinae
Genus *Brachidontes* Swainson, 1840

Type species: (by monotypy) *Modiola sulcata* Lamarck.

Subgenus *Brachidontes* s.s.
Brachidontes (Brachidontes) cowlitzensis (Weaver and Palmer)

Modiola ornata **Gabb, 1864; 184, 185, pl. 24, fig. 166.**
Modiolus (Brachydontes) cowlitzensis **Weaver and Palmer, 1922:16, 17, pl. 9, fig. 19.**
Brachidontes cowlitzensis? (Weaver and Palmer). Stewart, 1930:100, pl. 8, fig. 12.
Brachidontes cowlitzensis (Weaver and Palmer). Turner, 1938:45, 46, pl. 6, figs. 7, 8.
Volsella (Brachidontes) cowlitzensis **(Weaver and Palmer). Weaver, 1942:113, 114, pl. 26, fig. 4.**

Hypotypes: **UCR 4706/25, locality 4706; UCR 4721/15 and 4726/35, localities 4721 and 4726, respectively.**

Local occurrence: ***Ectinochilus supraplicatus*** **and *E. canalifer* faunas, Juncal and Matilija formations.**

Provincial Range: Late Paleocene ("Meganos Stage") to late Eocene (***Turritella variata lorenzana*** Zone of Kleinpell and Weaver, 1963).

Superfamily PTERIACEA
Family Pteriidae
Genus *Pteria* Scopoli, 1777

Type species: **(by subsequent designation, Kennard, Salisbury and Woodward, 1931) *Mytilus hirundo* Linné.**

Pteria pellucida (Gabb)

(Pl. 1, fig. 10)

Avicula pellucida **Gabb, 1864:186, 187, pl. 25, fig. 172.**
Pteria pellucida **(Gabb). Anderson and Hanna, 1925:188, 189, pl. 1, fig. 1. Vokes, 1939:50, 51, pl. 2, figs. 1, 4, 7, 8. Kleinpell and Weaver, 1963:197, pl. 29, fig. 5.**

Hypotype: **UCR 4714/6, locality 4714.**

Local occurrence: ***Ectinochilus canalifer*** **fauna, Matilija Sandstone.**

Provincial Range: **Middle Eocene ("Domengine Stage") to late Eocene (*Turritella schencki delaguerrae* Zone of Kleinpell and Weaver, 1963).**

Family Isognomonidae
Genus *Isognomon* Lightfoot, 1786

Type species: **(by monotypy) *Ostrea perna* Linné.**

Isognomon n. sp.?

(Pl. 2, fig. 6)

Hypotype: **UCR 4752/61, locality 4752, *Turritella uvasana applinae* fauna, Juncal Formation.**

Five poorly preserved specimens of a large *Isognomon* were collected from locality 4752. They represent the first record of this genus in the Eocene of the Pacific Coast and, therefore, probably represent a new species. The small size of the sample and poor preservation of the specimens, however, do not justify the naming of a new species at this time. The description of the figured specimen is as follows: shell subquadrate in outline, about 6 in high and 4 in long, moderately convex, equivalve, inequilateral; umbo situated at anterior end of hinge margin, produced and directed anteriorly; anterior margin of shell nearly straight, becoming concave near beak and forming an acute angle with the hinge margin; posterior and ventral margins of shell evenly rounded; narrow byssal gape present below beaks; hinge area with at least 11 narrow, subequal ligament grooves; surface of shell concentrically lamellose.

"Perna" goniglensis Hanna (1927:275, pl. 27, figs. 12–14) and *"Pedalion" joaquinensis* Vokes (1939:52, 53, pl. 2, figs. 3, 6, 10, 12, 15) from the middle Eocene ("Domengine Stage") of California are distinguished from *Isognomon* n. sp.? by their much smaller size and by the presence of only 3–5 ligament grooves on the hinge area. These species appear to be referable to *Pachyperna* Oppenheim (see Cox et al., 1969, part N, 1:326) of the European middle Eocene.

Family MALLEIDAE
Genus *Nayadina* Munier-Chalmas, 1864

Type species: (by monotypy) *Nayadina heberti* Munier-Chalmas.

Subgenus *Exputens* Clark, 1934

Type species: (by subsequent designation, Vokes, 1939) *Exputens llajasensis* Clark.

Nayadina (*Exputens*) *llajasensis* (Clark)

(Pl. 1, fig. 9)

Exputens llajasensis Clark, 1934:270, 271, pl. 37, figs. 11–18.

Hypotype: UCR 4659/81, locality 4659.

Local occurrence: Turritella uvasana infera fauna, Juncal Formation.

Provincial Range: Early Eocene ("Capay Stage") to middle Eocene ("Domengine Stage").

A few specimens referable to this species were collected from locality 4659. It has previously been reported only from its type locality in the Llajas Formation in the Simi Valley, Ventura County, California.

Superfamily OSTREACEA
Family OSTREIDAE
Genus *Ostrea* Linné, 1758

Type species: (by subsequent designation, Children, 1823) *Ostrea edulis* Linné.

Ostrea idriaensis Gabb

Ostrea idriaensis Gabb, 1869:203, pl. 33, figs. 103*b, c, d;* pl. 34, figs. 103, 103*a*. Hanna, 1927:276, pl. 30, figs. 1, 2; pl. 31, figs. 3, 4. Stewart, 1930:126, 127, pl. 8, fig. 3; pl. 17, fig. 1. Vokes, 1935*a*:291–304, pls. 22–24; 1939: 54. Turner, 1938:46, pl. 6, fig. 9.

[non] *Ostrea haleyi* Hertlein, 1933: 277–282, pl. 18, figs. 5, 6.

Hypotype: UCR 4747/1, locality 4747.

Local occurrence: Ectinochilus supraplicatus and *E. canalifer* faunas, Juncal and Matilija formations.

Provincial Range: Early Eocene ("Capay Stage") to late Eocene (*Turritella schencki delaguerrae* Zone of Kleinpell and Weaver, 1963).

Genus *Odontogryphaea* Ihering, 1903

Type species: (by original designation) *Gryphaea consors* var. *rostrigera* Ihering.

Odontogryphaea? *haleyi* (Hertlein)

(Pl. 1, figs. 11–13)

Ostrea haleyi Hertlein, 1933:277–282, pl. 18, figs. 5, 6.
Hypotypes: UCR 4670/32, 4670/33, 4670/34, locality 4670.
Local occurrence: *Turritella uvasana infera* fauna, Juncal Formation.
Provincial Range: Early Eocene ("Capay Stage").

Disarticulated shells of this species, all left valves, were collected from several localities in the lower part of the mudstone facies of the Juncal Formation.

Vokes (1935*a*:291) considered *Odontogryphaea*? *haleyi* to be only a gryphaeoid variant of *Ostrea idriaensis* Gabb. All specimens of *O.*? *haleyi* collected during this study, however, have a gryphaeoid shape and it appears to be characteristic of this species. *O.*? *haleyi* is also distinguished from *Ostrea idriaensis* by smaller average size (the largest specimen collected has a height of 45 mm, as compared with an average height of 92 mm reported by Vokes, 1935*a*:292, for *O. idriaensis*), finer concentric ornamentation, lack of radial ribbing, and by the presence of a distinct posterior radial sulcus and posterior auricle on the left valve. Distinct pits, or *anachomata* (Stenzel, 1971:990), are present along the commissural shelf near the hinge in the left valve of *O.*? *haleyi.*

The generic position of this species is uncertain. Although it has a gryphaeoid shape, the presence of *anachomata* indicates that it is not a true *Gryphaea* (see Stenzel, 1971:1097–1099). It is similar in appearance to *Odontogryphaea* Ihering and is questionably assigned to this genus, although the specimens at hand are too poorly preserved to be able to determine whether the terebratuloid fold in the valve commissure characteristic of *Odontogryphaea* is present.

Superfamily LUCINACEA
Family Lucinidae
Subfamily Lucininae
Genus *Linga* de Gregorio, 1884

Tpye species: (by subsequent designation, Sacco, 1889) *Lucina columbella* Lamarck.

Linga cf. *L. taffana* (Dickerson)

Hypotype: UCR 4679/26, locality 4679, *Turritella uvasana applinae* fauna, Juncal Formation.

A single poorly preserved specimen comparable with this species was collected from locality 4679.

Subfamily Milthinae
Genus *Claibornites* Stewart, 1930

Type species: (by original designation) *Lucina rotunda* Lea.

Claibornites diegoensis (Dickerson)

(Pl. 1, fig. 15)

Lucina diegoensis Dickerson, 1916:484, pl. 37, figs. 1*a*, 1*b*.
Hypotype: UCR 4752/41, locality 4752.

Local occurrence: Turritella uvasana applinae fauna, Juncal Formation.
Provincial Range: Middle Eocene ("Domengine Stage").

This species is characterized by its compressed, discoidal shape; obsolete posterior areas; completely submerged ligament; and deeply excavated lunule that partly obscures the anterior cardinal tooth. These features indicate a close relationship with the genera *Claibornites* Stewart and *Saxolucina* Stewart (1930: 184) (see also Cox et al., 1969, part N, 2:502, 505). It is referred to *Claibornites* because of the presence of a strong right anterior lateral tooth. The lateral teeth are obsolete in *Saxolucina.*

Genus *Miltha* H. and A. Adams, 1857

Type species: (by original designation) *Lucina childreni* Gray.

Miltha cf. *M. Packi* (Dickerson)

Hypotype: UCR 4665/2, locality 4665; UCR 4700/31, locality 4700.
Local occurrence: *Turritella uvasana infera* and *T. uvasana applinae* faunas, Juncal Formation.

Large, poorly preserved specimens comparable with this species were collected from localities 4655, 4665, 4668, and 4700 in the Juncal Formation.

Family Ungulinidae
Genus *Diplodonta* Bronn, 1831

Type species: (by subsequent designation, Herrmannsen, 1846) *Venus lupinus* Brocchi.

Diplodonta unisulcatus (Vokes)

Taras unisulcatus Vokes, 1939: 74, pl. 10, figs. 4, 7, 10.
Hypotype: UCR 4679/20, locality 4679.
Local occurrence: Turritella uvasana applinae fauna, Juncal Formation.
Provincial Range: Middle Eocene ("Domengine Stage").

Superfamily CARDITACEA
Family Carditidae
Subfamily Venericardiinae
Genus *Venericardia* Lamarck, 1801

Type species: (by subsequent designation, Schmidt, 1818) *Venericardia imbricata* Lamarck.

Subgenus *Pacificor* Verastegui, 1953

Type species: (by original designation) *V.* (*P.*) *mulleri* Verastegui.

Venericardia (*Pacificor*) *hornii* (Gabb) s.s.

(Pl. 3, fig. 3)

Cardita hornii Gabb, 1864: 174, pl. 24, fig. 157; 1869:187, 188, pl. 30, figs. 83, 83*a*.
Venericardia hornii (Gabb). Hanna, 1925:286, 287, pl. 37, fig. 5; pl. 38, fig. 1; pl. 39, figs. 1, 2. Anderson and Hanna, 1925:174, pl. 4, fig. 1.
Venericardia (*Venericor*) *hornii* (Gabb). Stewart, 1930:165–168, pl. 11, fig. 1.
Venericardia (*Pacificor*) *hornii* (Gabb). Verastegui, 1953:33–35, pl. 18, figs., 1–7; pl. 19, fig. 7.
Hypotype: UCR 4745/1, locality 4745.
Local occurrence: Ectinochilus canalifer fauna, Matilija Sandstone and Coldwater Sandstone (cf. *V. hornii* s.s.).
Provincial Range: Late Eocene ("Tejon Stage").

Typical *Venericardia hornii* is characterized by an obliquely ovate outline, a pointed to sharply rounded posteroventral margin, and 20–22 rounded radial ribs. It occurs only in the *Ectinochilus canalifer* fauna in the Pine Mountain section.

Venericardia (*Pacificor*) *hornii* (Gabb) cf. subsp. *calafia* Stewart

(Pl. 4, fig. 1)

Hypotype: UCR 4684/6, locality 4684, *Turritella uvasana applinae* fauna, Juncal Formation.
Local occurrence: Turritella uvasana applinae fauna, Juncal Formation.

Poorly preserved specimens that appear to represent this subspecies were collected from several localities in the *Turritella uvasana applinae* fauna. *V. hornii calafia* is characterized by its circular to subquadrate outline and by the presence of 24 radial ribs (see Stewart, 1930:169; Verastegui, 1953:28–30).

Verastegui (1953:28) considered this form to be a distinct species. I prefer to follow Stewart (1930:169) and regard it as a subspecies of *V. hornii,* at least until its range of variation is better known.

Venericardia (*Pacificor*) *hornii lutmani* Turner

(Pl. 1, fig. 16; pl. 2, fig. 8)

Venericardia hornii (Gabb) *lutmani* Turner, 1938:50, pl. 13, fig. 4; pl. 14, fig. 2. Weaver, 1942: 135, pl. 28, fig. 1; pl. 32, fig. 1.
Venericardia (*Pacificor*) *lutmani* Turner. Verastegui, 1953:26, 27, pl. 7, figs. 3–5; pl. 8, fig. 8.
Hypotype: UCR 4662/40, locality 4662.
Local occurrence: Turritella uvasana infera fauna, Juncal Formation.
Provincial Range: Late Paleocene ("Meganos Stage") to early Eocene ("Capay Stage").

This subspecies is similar in outline to *V. hornii calafia,* but is distinguished by a larger number (27–30) of radial ribs. As suggested by Turner (1938:50), *V. hornii lutmani* probably evolved into *V. hornii calafia* by reduction in the number of ribs.

Genus *Glyptoactis* Stewart, 1930

Type species: (by original designation) *Venericardia hadra* Dall.

Glyptoactis domenginica (Vokes)

Venericardia (*Glyptoactis?*) *domenginica* Vokes, 1939:66, pl. 5, figs. 7–9.
Venericardia (*Glyptoactis*) *domenginica* Vokes. Verastegui, 1953:43, pl. 13, fig. 1.
Hypotypes: UCR 4675/6, locality 4675; UCR 4679/21, locality 4679; UCR 4707/11, locality 4707.
Local occurrence: Turritella uvasana applinae and *Ectinochilus supraplicatus* faunas, Juncal Formation.
Provincial Range: Middle Eocene ("Domengine" and "Transition" "Stages").

Superfamily CRASSATELLACEA
Family CRASSATELLIDAE
Subfamily CRASSATELLINAE
Genus *Crassatella* Lamarck, 1799

Type species: (by subsequent designation, Schmidt, 1818) *Mactra cygnaea* Lamarck.

Crassatella mulates (Hanna)

Crassatellites mulates Hanna, 1927:282, pl. 34, figs. 2, 3, 6, 7, 9.
Crassatella mulates (Hanna). Vokes, 1939:63, pl. 4, figs. 14, 15.

Hypotypes: UCR 4690/18 and 4690/181, locality 4690.
Local occurrence: Turritella uvasana applinae fauna, Juncal Formation.
Provincial Range: Middle Eocene ("Domengine Stage") to (?) late Eocene ("Tejon Stage").

This species is characterized by the straight posterior margin of the shell and by the prominent posterior angulation of the shell surface. Two specimens, one right valve and one left valve, were collected from locality 4690.

Crassatella uvasana Conrad s.s.

Crassatella uvasana Conrad, 1855:9; 1857, pl. 2, fig. 5. Gabb, 1864:214, pl. 32, fig. 284. Stewart, 1930:141–143, pl. 12, fig. 9.
Crassatellites uvasanus (Conrad). Anderson and Hanna, 1925:172, 173, pl. 4, figs. 2, 3; text fig. 7.
Hypotype: UCR 4741/13, locality 4741.
Local occurrence: Ectinochilus canalifer fauna, Matilija Sandstone.
Provincial Range: Late Eocene ("Tejon Stage").

This species is characterized by its large, thick, robust shell; nearly equilateral, triangular shape; inconspicuous umbonal ridge; broad, prominent beaks; and low concentric undulations on the umbones. The beaks are lower on the typical subspecies than on *C. uvasana semidentata.*

Crassatella uvasana semidentata (Cooper)

Astarte semidentata Cooper, 1894:48, pl. 3, figs. 44, 45.
Crassatellites grandis (Gabb). Arnold, 1909:13, pl. 2, figs. 10, 10*a;* pl. 3, fig. 14.
Crassatellites mathewsonii (Gabb). Dickerson, 1916:430, 444, pl. 36, figs. 9*a,* 9*b.*
Crassatella uvasana (Conrad) *semidentata* (Cooper). Vokes, 1939:64, pl. 4, figs. 4, 6, 8, 10, 12.
Hypotypes: UCR 4672/3, locality 4672; UCR 4679/10, locality 4679; UCR 4707/13, locality 4707.
Local occurrence: Turritella uvasana applinae and *Ectinochilus supraplicatus* faunas, Juncal Formation.
Provincial Range: Middle Eocene ("Domengine" and "Transition" "Stages").

This subspecies has its highest known occurrence within the "Transition Stage" and is apparently ancestral to *C. uvasana* s.s. of the "Tejon Stage." It is distinguished from the typical subspecies by higher, markedly attenuated, umbos.

Superfamily CARDIACEA
Family CARDIIDAE
Subfamily CARDIINAE
Genus *Acanthocardia* Gray, 1851

Type species: (by subsequent designation, Stoliczka, 1870) *Cardium aculeatum* Linné.

Subgenus *Schedocardia* Stewart, 1930

Type species: (by original designation) *Cardium hatchetigbeense* Aldrich.

Acanthocardia (*Schedocardia*) *brewerii* (Gabb)

(Pl. 1, fig. 17)

Cardium brewerii Gabb, 1864:173, pl. 24, fig. 155. Anderson and Hanna, 1925:165, 166, pl. 1, fig. 3.
Plagiocardium (*Schedocardia*) *brewerii* (Gabb). Stewart, 1930:256–258, pl. 12, fig. 6. Turner, 1938:52, 53, pl. 9, figs. 6, 7. Vokes, 1939:75, pl. 11, figs. 1–4.
Cardium (*Trachycardium*) *brewerii brewerii* (Gabb). Kleinpell and Weaver, 1963:201, pl. 34, figs. 1, 2.

Hypotype: UCR 4741/4, locality 4741.

Local occurrence: Turritella uvasana applinae and *Ectinochilus canalifer* faunas, Juncal and Matilija formations.

Provincial Range: Middle Eocene ("Domengine Stage") to late Eocene (*Turritella schencki delaguerrae* Zone of Kleinpell and Weaver, 1963).

This species is very abundant in the *E. canalifer* fauna. A few specimens were also collected from the *Turritella uvasana applinae* fauna.

Subfamily PROTOCARDIINAE

Genus *Nemocardium* Meek, 1876

Type species: (by subsequent designations, Sacco, 1899) *Cardium semiasperum Deshayes.*

Subgenus *Nemocardium* s.s.

Nemocardium (*Nemocardium*) *linteum* (Conrad)

Cardium linteum Conrad, 1855:3, 9; 1857, pl. 2, fig. 1. Anderson and Hanna, 1925:166, 167, pl. 3, fig. 3.

Cardium cooperii Gabb, 1864:172, pl. 24, figs. 154, 154*a*.

Nemocardium linteum (Conrad). Stewart, 1930:275–277, pl. 8, fig. 6. Turner, 1938:52, pl. 10, fig. 10. Vokes, 1939:76, 77, pl. 11, figs. 6, 9.

Cardium (*Nemocardium*) *linteum* Conrad. Kleinpell and Weaver, 1963:202, pl. 34, fig. 4.

Hypotypes: UCR 4682/8, locality 4682; UCR 4707/9, locality 4707; UCR 4723/9, locality 4723.

Local occurrence: Turritella uvasana applinae, Ectinochilus supraplicatus, and *E. canalifer* faunas, Juncal and Matilija formations.

Provincial Range: Middle Eocene ("Domengine Stage") to late Eocene ("Tejon Stage")

Superfamily MACTRACEA

Family MACTRIDAE

Subfamily MACTRINAE

Genus *Spisula* Gray, 1837

Type species: (by subsequent designation, Gray, 1837) *"Mactra solida* Montague" (=*Cardium solidum* Linné, 1758).

Spisula bisculpturata Anderson and Hanna

(Pl. 2, fig. 5)

Spisula bisculpturata Anderson and Hanna, 1925:149–151, pl. 3, fig. 7.

Hypotypes: UCR 4719/10, locality 4719; UCR 4743/701, locality 4743.

Local occurrence: Ectinochilus canalifer fauna, Juncal and Matilija formations.

Provincial Range: Late Eocene ("Tejon Stage").

This species is very abundant in the *Ectinochilus canalifer* fauna.

Superfamily SOLENACEA

Family SOLENIDAE

Genus *Solena* Mörch, 1853

Type species: (by subsequent designation, Stoliczka, 1871) *Solen obliquus* Spengler.

Subgenus *Eosolen* Stewart, 1930

Type species: (by original designation) *Solen plagiaulax* Cossmann.

Solena (*Eosolen*) *coosensis* Turner

(Pl. 2, fig. 1)

Solen novacula Anderson and Hanna. Hanna, 1927:294, pl. 43, fig. 1.

[non] *Solen novacula* Anderson and Hanna, 1925:147, pl. 6, fig. 9.

Solena* (*Eosolen*) *coosensis Turner, 1938:62, pl. 9, figs. 1, 2. Vokes, 1939:96, pl. 15, fig. 5.
Hypotype: UCR 4676/22, locality 4676.
Local occurrence: Turritella uvasana applinae fauna, Juncal Formation.
Provincial Range: Early Eocene ("Capay Stage") to middle Eocene ("Domengine Stage").

Solena (*Eosolen*) *subverticala* Vokes

Solena* (*Eosolen*) *subverticala Vokes, 1939:96, pl. 15, fig. 8.
Hypotype: UCR 4683/13, locality 4683.
Local occurrence: Turritella uvasana applinae fauna, Juncal Formation.
Provincial Range: Early Eocene ("Capay Stage") to middle Eocene ("Domengine Stage").

Superfamily TELLINACEA
Family Tellinidae
Subfamily Tellininae
Genus *Tellina* Linné, 1758

Type species: (by subsequent designation, Children, 1823) *Tellina radiata* Linné.

Tellina castacana Anderson and Hanna

(Pl. 2, fig. 2)

Tellina castacana Anderson and Hanna, 1925:153, pl. 2, fig. 13.
Hypotype: UCR 4743/502, locality 4743.
Local occurrence: Ectinochilus canalifer fauna, Juncal and Matilija formations.
Provincial Range: Late Eocene ("Tejon Stage").

Tellina lebecki Anderson and Hanna

Tellina lebecki Anderson and Hanna, 1925:154, pl. 3, fig. 10.
Hypotype: UCR 4741/16, locality 4741.
Local occurrence: Ectinochilus canalifer fauna, Matilija Sandstone.
Provincial Range: Late Eocene ("Tejon Stage").

Four specimens referable to this species were collected from the *E. canalifer* fauna. *Tellina lebecki* resembles *T. tehachapi* and *T. soledadensis,* but is distinguished by a lower and more elongate shell and by the lack of radial ribbing.

Tellina soledadensis Hanna

(Pl. 2, fig. 7)

Tellina soledadensis Hanna, 1927:291, pl. 42, figs. 1, 2, 5. Turner, 1938:60, 61, pl. 7, fig. 5. Vokes, 1939:90, pl. 14, fig. 13. Weaver, 1942:196, 197, pl. 48, fig. 1.
Hypotypes: UCR 4678/16, locality 4678; UCR 4703/1, locality 4703; UCR 4708/3, locality 4708.
Local occurrence: Turritella uvasana applinae, and *Ectinochilus supraplicatus* faunas, Juncal Formation.
Provincial Range: Early Eocene ("Capay Stage") to late middle Eocene ("Transition Stage").

This species has its highest known occurrence within the "Transition Stage" and is apparently ancestral to *Tellina tehachapi* of the "Tejon Stage."

Tellina tehachapi Anderson and Hanna

Tellina remondii Gabb, 1869:182, pl. 29, fig. 71; not Gabb, 1864:156, pl. 22, fig. 132.
Tellina tehachapi Anderson and Hanna, 1925:155, pl. 6, figs. 5, 6. Kleinpell and Weaver, 1963: 207, pl. 37, fig. 8.
Hypotypes: UCR 4714/7, locality 4714; UCR 4723/5, locality 4723; UCR 4741/24, locality 4741.
Local occurrence: Ectinochilus canalifer fauna, Matilija Sandstone.

Provincial Range: Late Eocene ("Tejon Stage" and *Turritella schencki delaguerrae* Zone of Kleinpell and Weaver, 1963).

Poorly preserved specimens of this species were collected from several localities in the Matilija Sandstone. It is distinguished from *Tellina soledadensis,* its apparent ancestor, by more broadly spaced concentric lamellae.

Tellina n. sp.? aff. *T. townsendensis* Clark

(Pl. 2, fig. 3)

Hypotype: UCR 4743/6, locality 4743, *Ectinochilus canalifer* fauna, Matilija Sandstone.

Specimens of a *Tellina* similar to *T. townsendensis* Clark (1925:94, pl. 12, figs. 11, 12) from the Oligocene Gries Ranch Beds of Washington, but distinguished by a slightly lower and more elongate shell, were collected from a number of localities in the Matilija Sandstone. They may represent a new species.

Subfamily MACOMINAE

Genus *Macoma* Leach, 1819

Type species: (by monotypy) *Macoma tenera* Leach, 1819 (= *Tellina calcarea* Gmelin, 1791).

Macoma viticola Anderson and Hanna

(Pl. 2, fig. 4)

Macoma viticola Anderson and Hanna, 1925:157, pl. 2, fig. 12.
Hypotype: UCR 4726/12, locality 4726.
Local occurrence: Ectinochilus canalifer fauna, Juncal and Matilija formations.
Provincial Range: Late Eocene ("Tejon Stage").

This species is very abundant in the *E. canalifer* fauna.

Family PSAMMOBIIDAE

Subfamily PSAMMOBIINAE

Genus *Gari* Schumacher, 1817

Type species: (pending decision by the Intern. Comm. Zool. Nomen.) *Gari vulgaris* Schumacher, 1817 (= *Solen amethystus* Wood, 1815).

Gari eoundulata Vokes

Gari eoundulata Vokes, 1939:93, 94, pl. 14, figs. 23, 24.
Hypotype: UCR 4672/6, locality 4672.
Local occurrence: Turritella uvasana applinae fauna, Juncal Formation.
Provincial Range: Middle Eocene ("Domengine Stage").

A single specimen referable to this species was collected from locality 4672.

Gari hornii (Gabb) s.s.

Tellina hornii Gabb, 1864:160, 161, pl. 30, fig. 244.
Psammobia hornii (Gabb). Anderson and Hanna, 1925:151, 152, pl. 3, fig. 6; pl. 9, fig. 16.
Gari hornii (Gabb). Stewart, 1930:282, 283, pl. 12, fig. 2. Weaver, 1942:218, pl. 50, fig. 13; pl. 51, fig. 3. Kleinpell and Weaver, 1963:207, pl. 38, fig. 3.
Hypotype: UCR 4715/12, locality 4715.
Local occurrence: Ectinochilus canalifer fauna, Juncal and Matilija formations.
Provincial Range: Late middle Eocene ("Transition Stage") to late Eocene (*Turritella schencki delaguerrae* Zone of Kleinpell and Weaver, 1963).

Superfamily CORBICULACEA
Family CORBICULIDAE
Genus *Corbicula* Mergele von Mühlfeld, 1811

Type species: (by subsequent designation, Intern. Comm. Zool. Nomen., 1955) *Tellina fluminalis* Müller.

Corbicula n. sp.? aff. *C. williamsoni* Anderson and Hanna

Hypotype: UCR 4663/31, locality 4663, *Ectinochilus canalifer* fauna, Coldwater Sandstone.

Three poorly preserved specimens of *Corbicula* collected from locality 4663 resemble *C. williamsoni* Anderson and Hanna (1925:164, 165, pl. 1, fig. 4; pl. 3, fig. 2) from the type Tejon Formation, but are distinguished by higher beaks and a more prominent umbonal ridge. They may represent a new species.

Superfamily VENERACEA
Family VENERIDAE
Subfamily MERETRICINAE
Genus *Tivelina* Cossmann, 1886

Type species: (by subsequent designation, Crosse, 1886) *Cytherea rustica* Deshayes.

Tivelina cf. *T. vaderensis* (Dickerson)

Hypotype: UCR 4696/4, locality 4696, *Ectinochilus supraplicatus* fauna, Juncal Formation.

Several poorly preserved specimens of *Tivelina* comparable with *T. vaderensis* (Dickerson, 1915:54, pl. 3, figs. 5*a–c*) were collected from locality 4696.

Subfamily PITARINAE
Genus *Pitar* Römer, 1857

Type species: (by monotypy) *Venus tumens* Gmelin.

Subgenus *Pitar* s.s.
Pitar (*Pitar*) *californianus* (Conrad) s.s.

Meretrix californiana Conrad, 1855:9; 1857, pl. 2, fig. 4.
Pitaria californiana (Conrad). Anderson and Hanna, 1925:159, pl. 5, figs. 1, 2.
Pitar californiana (Conrad). Turner, 1938:53, pl. 12, figs. 4, 5. Weaver, 1942:177, 178, pl. 40, figs. 10, 13; pl. 41, figs. 15–19; pl. 47, figs. 6, 12.
Pitar californianus (Conrad). Keen and Bentson, 1944:101.
Hypotype: UCR 4741/10, locality 4741.
Local occurrence: Ectinochilus canalifer fauna, Matilija Sandstone.
Provincial Range: Late Eocene ("Tejon Stage").

A single well-preserved specimen of this species was collected from locality 4741.

Pitar (*Pitar*) cf. *P.* (*P.*) *palmeri* (Clark and Woodford)

Hypotype: UCR 4664/61, locality 4664, *Turritella uvasana infera* fauna, Juncal Formation.

Several large, poorly preserved specimens comparable with *P. palmeri* (Clark and Woodford, 1927: 96, pl. 16, figs. 7–9) were collected from localities 4664 and 4670 in the *Turritella uvasana infera* fauna.

Pitar (*Pitar*) cf. *P.* (*P.*) *lascrucensis* Kleinpell and Weaver

(Pl. 4, fig. 2)

Hypotype: UCR 4732/101, locality 4732.

Local occurrence: Ectinochilus canalifer fauna, Matilija Sandstone.

Poorly preserved specimens of a *Pitar* s.s. comparable with *P. lascrucensis* Kleinpell and Weaver (1963:205, pl. 36, figs. 2, 3, 5) were collected from several localities in the upper part of the Matilija Sandstone near Pine Mountain Lodge. The shell of the Pine Mountain specimens is ornamented with fine concentric growth lines and faint radial lines. The pallial sinus is deep, pointed, and ascending. The hinge is not exposed.

Subgenus *Calpitaria* Jukes-Browne, 1908

Type species: (by original designation) *Cytherea sulcataria* Deshayes.

Pitar (*Calpitaria*) *kelloggi* (Hanna)

Antigona kelloggi Hanna, 1927:286, pl. 38, figs. 7, 10–12.
Pitar (*Calpitaria*) *uvasanus kelloggi* (Hanna). Stewart, 1946, table 1.
Hypotype: UCR 4679/6, locality 4679.
Local occurrence: Turritella uvasana applinae fauna, Juncal Formation.
Provincial Range: Middle Eocene ("Domengine Stage").

A few specimens referable to this species were collected from the *Turritella uvasana applinae* fauna. It is closely related to *P.* (*C.*) *uvasanus* (Conrad) s.s., but is distinguished by its lower altitude, greater length and more closely spaced concentric lamellae.

Pitar (*Calpitaria*) *uvasanus* (Conrad) s.s.

Meretrix uvasana Conrad, 1855:9; 1857, pl. 2, fig. 3. Gabb, 1864:163, pl. 30, fig. 248.
Meretrix tejonensis Dickerson, 1915:42, 53 [in part; not pl. 3, figs. 2*a*–2*b*]. [New name for "*Meretrix uvasana* Conrad" of Gabb, 1864, not Conrad.]
Pitar (*Calpitaria*) *uvasanus* (Conrad). Stewart, 1930:235, 236, pl. 12, fig. 7.
Hypotypes: UCR 4707/14, locality 4707; UCR 4715/11, locality 4715; UCR 4738/11, locality 4738.
Local occurrence: Ectinochilus supraplicatus and *E. canalifer* faunas, Juncal and Matilija formations.
Provincial Range: Late middle Eocene ("Transition Stage") to late Eocene ("Tejon Stage").

Subgenus *Lamelliconcha* Dall, 1902

Type species: (by original designation) *Cytherea concinna* Sowerby.

Pitar (*Lamelliconcha*) *avenalensis* Vokes

(Pl. 3, Figs. 1, 4)

Pitar (*Lamelliconcha*) *avenalensis* Vokes, 1939, p. 86, pl. 13, figs. 4, 5, 8.
Hypotypes: UCR 4679/30 and 4679/31, locality 4679.
Local occurrence: T. uvasana applinae fauna, Juncal Formation.
Provincial Range: Middle Eocene ("Domengine Stage").

This small distinctive species is abundant at several localities in the *Turritella uvasana applinae* fauna.

Pitar (Lamelliconcha) dickersoni Givens, n.s.

(Pl. 3, fig. 10)

Meretrix tejonensis Dickerson, 1915:53 [in part], pl. 3, figs. 2*a*, 2*b*.
Pitaria tejonensis (Dickerson). Anderson and Hanna, 1925:160, pl. 3, fig. 5.
Holotype: California Academy of Sciences no. 262, locality CAS 244, type Tejon Formation.
Paratype: California Academy of Sciences No. 263, locality CAS 244, type Tejon Formation.

Hypotypes: UCR 4719/121, locality 4719; UCR 4721/171, locality 4721.
Local occurrence: Ectinochilus canalifer fauna, Matilija Sandstone.
Provincial Range: Late Eocene ("Tejon Stage"); questionably identified from the *Turritella schencki delaguerrae* Zone in western Santa Barbara County (Kleinpell and Weaver, 1963:205, pl. 36, fig. 9).

The new name *Pitar* (*Lamelliconcha*) *dickersoni* is proposed for the species originally described and figured by Dickerson (1915) as *Meretrix tejonensis.* As pointed out by Stewart (1930:236), Dickerson's name cannot be used for this species because it is an objective synonym of *Pitar uvasanus* (Conrad).

Pitar dickersoni resembles *P. uvasanus* but is much smaller in size (the largest specimen of *P. dickersoni* collected during this study has a height of 17.5 mm and length of 21.0 mm as compared with the height and length of 44.9 mm and 55.7 mm, respectively, of the neotype of *P. uvasanus*), higher in proportion to length, has a more prominent umbo, and has thicker and more closely spaced concentric lamellae on the surface of the valves. The hinge of the holotype of *P. dickersoni,* a left valve, has been figured by Anderson and Hanna (1925, pl. 3, fig. 5). It consists of a strong triangular 2*b* joined to a thin 2*a,* a 4*b* which is parallel to and confluent with the nympth, and a strong anterior lateral tooth. The species is referred to the subgenus *Lamelliconcha* because of the strong concentric lamellae ornamenting the shell surface.

This species is abundant in the *Ectinochilus canalifer* fauna. It is known with certainty only from the "Tejon Stage," although it has also been questionably identified from the slightly younger *Turritella schencki delaguerrae* Zone (Kleinpell and Weaver, 1963).

Pitar (*Lamelliconcha*) *joaquinensis* Vokes

(Pl. 3, fig. 7)

Pitar (*Lamelliconcha*) *joaquinensis* Vokes, 1939:85, 86, pl. 13, figs. 9–12.
Hypotype: UCR 4673/201, locality 4673.
Local occurrence: Turritella uvasana applinae fauna, Juncal Formation.
Provincial Range: Middle Eocene ("Domengine Stage").

A few specimens referable to this species were collected from localities 4673 and 4699 in the *Turritella uvasana applinae* fauna. *P. joaquinensis* is characterized by broad, flat-topped, concentric lamellae that are separated by narrow, incised linear interspaces. *Pitar avenalensis* Vokes also has flat-topped concentric lamellae, but they are much narrower and this species is longer in proportion to height than *P. joaquinensis.*

The specimen of *Pitar joaquinensis* figured from the Pine Mountain section differs slightly from the type and paratypes of this species in having a lower, less inflated umbo. It more closely resembles the specimen figured by Stewart (1946, pl. 12, fig. 6) from the Avenal Sandstone on Reef Ridge.

Pitar (*Lamelliconcha*) *soledadensis* (Hanna)

(Pl. 3, fig. 2)

Pitaria soledadensis Hanna, 1927:288, 289, pl. 38, figs. 8, 9.
Hypotype: UCR 4705/1, locality 4705.
Local occurrence: E. supraplicatus fauna, Juncal Formation.
Provincial Range: Middle Eocene ("Domengine" and "Transition" "Stages").

One well-preserved and several poorly preserved specimens of this species were collected from the *Ectinochilus supraplicatus* fauna.

Genus *Callista* Poli, 1791

Type species: (by subsequent designation, Meek, 1876) *Venus chione* Linné.

Subgenus *Costacallista* Palmer, 1927

Type species: (by original designation) *Venus erycina* Linné.

Callista (*Costacallista*) *hornii* (Gabb)

(Pl. 3, fig. 6)

Meretrix hornii Gabb, 1864:164, pl. 23, fig. 144; 1869:185, pl. 30, fig. 78.
Antigona hornii (Gabb). Anderson and Hanna, 1925:158, 159, pl. 3, fig. 9; pl. 5, fig. 5.
Macrocallista (*Costacallista*?) *hornii* (Gabb). Stewart, 1930:242–244, pl. 12, fig. 8; pl. 17, fig. 7.
Macrocallista hornii (Gabb). Kleinpell and Weaver, 1963:203, pl. 35, figs. 5, 6.

Hypotype: UCR 4723/61, locality 4723.

Local occurrence: Ectinochilus canalifer fauna, Matilija Sandstone.

Provincial Range: Late middle Eocene ("Transition Stage") to late Eocene ("Tejon Stage").

This species is abundant at several localities in the Matilija Sandstone.

Subgenus *Macrocallista* Meek, 1876

Type species: (by monotypy) *Venus gigantea* Gmelin, 1791 (= *Venus nimbosa* Lightfoot, 1786).

Callista (*Macrocallista*) *andersoni* (Dickerson)

(Pl. 3, fig. 5)

Macrocallista (?) *andersoni* Dickerson, 1915:54, 55, pl. 4, figs. 1*a*–1*b*.
Macrocallista andersoni Dickerson. Anderson and Hanna, 1925:162, 163, pl. 3, fig. 8. Turner, 1938:56, pl. 10, fig. 17. Weaver, 1942:174, 175, pl. 40, figs. 4–6, 12, 14; pl. 46, fig. 14. Kleinpell and Weaver, 1963:203, pl. 35, figs. 3, 4.

Hypotype: UCR 4731/101, locality 4731.

Local occurrence: Ectinochilus canalifer fauna, Matilija Sandstone.

Provincial Range: Late middle Eocene ("Transition Stage") to late Eocene ("Tejon Stage" and *Turritella schencki delaguerrae* Zone of Kleinpell and Weaver, 1963).

Subgenus *Microcallista* Stewart, 1930

Type species: (by original designation) *Cytherea proxima* Deshayes.

Callista (*Microcallista*) *conradiana* (Gabb) s.s.

Tapes conradiana Gabb, 1864:169, pl. 32, fig. 282.
Microcallista? conradiana (Gabb). Stewart, 1930:244, 245, pl. 12, fig. 3.
Macrocallista (*Costacallista*) *conradiana* (Gabb). Turner, 1938:55, 56, pl. 10, figs. 11–14. Vokes, 1939:80.
Microcallista (*Costacallista*) *conradiana* (Gabb). Weaver, 1942:172, 173, pl. 41, figs. 5, 6; pl. 46, fig. 11; pl. 104, fig. 13. Kleinpell and Weaver, 1963:203, pl. 35, fig. 7.

Hypotype: UCR 4738/9, locality 4738.

Local occurrence: Ectinochilus canalifer fauna, Matilija Sandstone.

Provincial Range: Late Eocene ("Tejon Stage" and *Turritella schencki delaguerrae* Zone of Kleinpell and Weaver, 1963).

Callista (*Microcallista*) cf. *C.* (*M.*) *tecolotensis* (Hanna)

Hypotype: UCR 4680/24, locality 4680.

Local occurrence: Turritella uvasana applinae fauna, Juncal Formation.

A single poorly preserved specimen comparable with Hanna's (1927:287, pl. 38, figs. 2, 4–6, 13) species was collected from locality 4680.

Genus *Callocardia* A. Adams, 1864

Type species: (by monotypy) *Callocardia guttata* A. Adams.

Subgenus *Nitidavenus* Vokes, 1939

Type species: (by original designation) *Cytherea nitida* Deshayes.

Callocardia (*Nitidavenus*) *conradi* (Dickerson)

Marcia (?) *conradi* Dickerson, 1916:484, 485, pl. 38, fig. 3.

Pitaria conradi (Dickerson). Clark and Woodford, 1927:95.

Nitidavenus conradi (Dickerson). Vokes, 1939:83, pl. 12, figs. 19–21.

Hypotypes: UCR 4657/2, locality 4657, UCR 4667/3, locality 4667; UCR 4670/2, locality 4670.

Local occurrence: Turritella uvasana infera fauna, Juncal Formation.

Provincial Range: Late Paleocene ("Meganos Stage") to early Eocene ("Capay Stage").

A few well-preserved and numerous poorly preserved specimens of this species were collected from several localities in the *Turritella uvasana infera* fauna. Most of numerous closely spaced concentric ribs.

Genus *Pelecyora* Dall, 1902

Type species: (by original designation) *Cytherea hatchetigbeensis* Aldrich.

Subgenus *Pelecyora* s.s.

Pelecyora gabbi (Arnold)

(Pl. 3, figs. 8, 11)

Meretrix gabbi Arnold, 1909:49, pl. 3, fig. 4. Arnold and Anderson, 1910:70, pl. 25, fig. 4.

Pelecyora gabbi (Arnold). Vokes, 1939:88, pl. 14, figs. 1–3, 5, 9.

Hypotypes: UCR 4683/701 and 4683/702, locality 4683.

Local occurrence: Turritella uvasana applinae fauna, Juncal Formation.

Provincial Range: Middle Eocene ("Domengine Stage").

Well-preserved specimens of this distinctive species were collected from several localities in the *Turritella uvansana applinae* fauna. It is characterized by its highly inflated shell, ovately trigonal outline, faint lunule, and ornamentation of numerous closely spaced concentric ribs.

Subfamily CYCLININAE

Genus *Cyclinella* Dall, 1902

Type species: (by original designation) *Dosinia tenuis* Recluz.

Cyclinella elevata (Gabb)

(Pl. 3, fig. 9)

Dosinia elevata Gabb, 1864:167, pl. 30, fig. 252. Anderson and Hanna, 1925:163, pl. 5, fig. 6.

Cyclinella elevata (Gabb). Stewart, 1930:227–229, pl. 12, fig. 1; pl. 17, fig. 8.

Hypotypes: UCR 4706/28, locality 4706; UCR 4726/21, locality 4726.

Local occurrence: Ectinochilus supraplicatus and *E. canalifer* faunas, Juncal and Matilija formations.

Provincial Range: Late middle Eocene ("Transition Stage") to late Eocene ("Tejon Stage").

Superfamily MYACEA
Family CORBULIDAE
Subfamily CORBULINAE
Genus *Corbula* Bruguiere, 1797

Type species: (by subsequent designation, Schmidt, 1818) *Corbula sulcata* Lamarck.

Subgenus *Caryocorbula* Gardner, 1926

Type species: (by original designation) *Corbula alabamiensis* Lea.

Corbula (*Caryocorbula*) *dickersoni* Weaver and Palmer
(Pl. 4, fig. 7)

Corbula dickersoni Weaver and Palmer, 1922:24, pl. 9, figs. 9, 10. Weaver, 1942:257, pl. 61, figs. 13, 16, 17, 20.

Corbula (*Caryocorbula*) *dickersoni* Weaver and Palmer. Vokes, 1939:98, pl. 16, figs. 1, 5, 9.

Hypotypes: UCR 4706/181, locality 4706; UCR 4707/32, locality 4707; UCR 4708/22, locality 4708.

Local occurrence: Ectinochilus supraplicatus fauna, Juncal Formation.

Provincial Range: Middle Eocene ("Domengine Stage") to late Eocene ("Tejon Stage").

This species resembles *Corbula hornii* Gabb, but is distinguished by a sharper posterior umbonal ridge and a more sharply truncated posterior margin of the shell. Occasional specimens of both species have faint radial lines crossing the concentric ribs on the posterior half of the shell. These radial lines are most prominent between the midline and the umbonal ridge. *C. dickersoni* is distinguished from *C. parilis* Gabb by its greater elongation.

Corbula (*Caryocorbula*) *hornii* Gabb
(Pl. 4, fig. 6)

Corbula hornii Gabb, 1864:149, 150, pl. 29, fig. 238; 1869:176, pl. 29, figs. 62, *a*, *b*. Dickerson, 1915, pl. 4 figs. 5*a*, 5*b*. Anderson and Hanna, 1925:148, pl. 2, figs. 3, 4. Stewart, 1930:287, 288, pl. 12, figs. 4, 5. Turner, 1938:66, pl. 8, figs. 16, 17. Kleinpell and Weaver, 1963:208, pl. 38, fig. 5.

Hypotypes: UCR 4717/301, locality 4717; UCR 4719/111, locality 4719.

Local occurrence: Ectinochilus canalifer fauna, Juncal and Matilija formations.

Provincial Range: Late Eocene ("Tejon Stage" and *Turritella schencki delaguerrae* Zone of Kleinpell and Weaver, 1963).

Corbula (*Caryocorbula*) *parilis* Gabb
(Pl. 4, fig. 9)

Corbula parilis Gabb, 1864:150, pl. 29, figs. 239, 239*a*. Arnold, 1909:106, pl. 2, fig. 2. Dickerson, 1915:84, pl. 4, fig. 8; 1916, pl. 40, fig. 10. Hanna, 1927:295, pl. 43, figs. 7–11, 13. Stewart, 1930:288, 289, pl. 3, fig. 5. Turner, 1938:65, 66, pl. 8, figs. 11–14.

Corbula (*Caryocorbula*) *parilis* Gabb. Vokes, 1939:99, pl. 16, figs. 2, 3, 6, 7, 10.

Hypotypes: UCR 4662/7, locality 4662; UCR 4680/101, locality 4680; UCR 4706/12, locality 4706.

Local occurrence: Turritella uvasana infera, T. uvasana applinae, and *Ectinochilus supraplicatus* faunas, Juncal Formation.

Provincial Range: Early Eocene ("Capay Stage") to late middle Eocene ("Transition Stage").

Subgenus *Cuneocorbula* Cossmann, 1886

Type species: (by subsequent designation, Dall, 1898) *Corbula biangulata* Deshayes.

Corbula (*Cuneocorbula*) *torreyensis* Hanna

Corbula torreyensis **Hanna, 1927:296, 297, pl. 44, figs. 6–10, 15, 16. Turner, 1938:66, pl. 8, figs. 6, 7.**

Cuneocorbula torreyensis **(Hanna). Clark and Vokes, 1936:875, pl. 1, figs. 9, 11. Vokes, 1939:101, 102, pl. 16, figs. 16, 20, 21.**

Hypotype: UCR 4697/2, locality 4697.

Local occurrence: Ectinochilus supraplicatus fauna, Juncal Formation.

Provincial Range: Early Eocene ("Capay Stage")? to late middle Eocene ("Transition Stage").

Several well-preserved specimens of this species were collected from locality 4697.

Superfamily PHOLADOMYACEA

Family PHOLADOMYIDAE

Genus *Pholadomya* G. B. Sowerby, 1823

Type species: (by subsequent designation, Gray, 1847) *Pholadomya candida* Sowerby.

subgenus *Pholadomya* s.s.

Pholadomya (Pholadomya) n. sp.

(Pl. 4, fig. 5)

Hypotype: UCR 4662/110, locality 4662, *Turritella uvasana infera* fauna, Juncal Formation.

Description: shell of medium size, elongate, subtrigonal, equivalve, strongly inequilateral, strongly inflated anteriorly, with a small posterior gape; beaks terminal, subangular, slightly elevated, orthogyrate; posterodorsal margin horizontal, anterior margin nearly straight and sloping downward from beaks at an angle of about 60 degrees, posterior and ventral margins broadly and evenly rounded; surface ornamented by concentric undulations of variable width and prominence, crossed by 12 to 15 tuberculate radial ribs on the central part of the shell.

Dimensions of hypotype: length, 33 mm; height, 22 mm; thickness (both valves), 18 mm.

Pholadomya n. sp. is represented by one well-preserved and seven poorly preserved, deformed, specimens, all collected from localities in the lower part of the mudstone facies of the Juncal Formation. Because of the poor preservation of the material, a name is not proposed for this species at this time.

Only two other species of *Pholadomya* have been described from the Lower Tertiary of the Pacific Coast: *P. nasuta* Gabb (1864), from the Paleocene ("Martinez Stage"); and *P. murrayensis* Hanna (1927), from the late Eocene ("Tejon Stage"). Both are readily distinguished from the new species by their broader, nearly vertical, anterior margin; narrower posterior margin; sloping, slightly concave dorsal margin; and more prominent, strongly elevated beaks which, although anterior, are not situated at the anterior extremity of the shell.

Superfamily PANDORACEA

Family PERIPLOMATIDAE

Genus *Periploma* Schumacher, 1817

Type species: (by monotypy) *Corbula margaritacea* Lamarck.

Subgenus?

Periploma cf. *P. stewartvillensis* Clark and Woodford

Hypotype: UCR 4685/15, locality 4685, *Turritella uvasana applinae* fauna, Juncal Formation.

A single poorly preserved, slightly crushed specimen comparable with Clark and Woodford's (1927:89, pl. 14, fig. 11) species was collected from locality 4685.

Periploma n. sp.?

(Pl. 4, fig. 8)

Hypotypes: UCR 4731/91 and 4731/92, locality 4731; UCR 4732/2, locality 4732.
Local occurrence: Ectinochilus canalifer fauna, Matilija Sandstone.

Abundant specimens of *Periploma,* most of them badly crushed, were collected from limestone concretions in silty mudstone near the top of the Matilija Sandstone. They resemble *P. stewartvillensis* in outline, but are more than twice as large and are more elongate anteriorly. They probably represent a new species. The shell is of moderate size for the genus, thin, elongate-subquadrate, slightly inequivalve, and subnacreous. The beaks are moderately prominent, opisthogyrate, and are situated about one-fifth the length of the shell from the posterior end. The anterodorsal margin is long and straight; the posterodorsal margin short and straight; the anterior and ventral margins broadly rounded; and the posterior margin slightly produced, subtruncate, and slightly depressed below the remainder of the shell. There are two shallow posterior grooves radiating from the umbo of the left valve and a corresponding low ridge on the right valve. The surface ornamentation consists of fine growth lines and low, broad, concentric undulations. Fine radial lines are also visible on some specimens. The hinge is not accessible.

Dimensions of figured specimen (UCR 4732/2): length, 50.8 mm; height, 32.8 mm; thickness (both valves), 15.5 mm.

Family THRACIIDAE

Genus *Thracia* Sowerby, 1823

Type species: (by subsequent designation, Anton, 1839) "*Thracia pubescens* Lamarck" (= *Mya pubescens* Pulteney, 1799).

Subgenus *Thracia* s.s.

Thracia (*Thracia*) *dilleri* Dall

(Pl. 4, fig. 4)

Thracia dilleri Dall, 1898:929, 1524, pl. 34, fig. 19. Weaver, 1942:119, pl. 29, fig. 3. Kleinpell and Weaver, 1963:208, pl. 38, figs. 6, 7, 9.

Hypotypes: UCR 4727/101, locality 4727; UCR 4731/81, locality 4731.

Local occurrence: Ectinochilus canalifer fauna, Matilija Sandstone.

Provincial Range: Late Eocene ("Tejon Stage" to *Turritella variata lorenzana* Zone of Kleinpell and Weaver, 1963).

Several poorly preserved specimens of this species were collected from UCR localities 4727, 4728, and 4731 in the upper part of the Matilija Sandstone.

Thracia (*Thracia*) *sorrentoensis* Hanna

Thracia sorrentoensis Hanna, 1927:280, pl. 33, fig. 1.

Hypotypes: UCR 4676/24, locality 4676; UCR 4688/8, locality 4688.

Local occurrence: T. uvasana applinae fauna, Juncal Formation.

Provincial Range: Middle Eocene ("Domengine Stage").

Two specimens were collected from the *T. uvasana applinae* fauna.

Superfamily POROMYACEA

Family CUSPIDARIIDAE

Genus *Cardiomya* A. Adams, 1843

Type species: (by monotypy) *Neaera gouldiana* Hinds.

Cardiomya israelskyi (Hanna)

(Pl. 4, fig. 3)

Cuspidaria israelskyi Hanna, 1927:280, pl. 34, figs. 1, 5.
Hypotypes: UCR 4680/3, locality 4680; UCR 4700/2, locality 4700.
Local occurrence: *T. uvasana applinae* fauna, Juncal Formation.
Provincial Range: Middle Eocene ("Domengine Stage").

Two specimens of this species were collected.

Class SCAPHOPODA
Family DENTALIIDAE
Genus *Dentalium* Linné, 1758

Type species: (by subsequent designation, Montfort, 1810) *Dentalium elephantinum* Linné.

Dentalium cooperii Gabb

Dentalium cooperii Gabb, 1864:139, pl. 21, fig. 100.
Hypotype: UCR 4706/311, locality 4706.
Local occurrence: *Turritella uvasana applinae* and *Ectinochilus supraplicatus* faunas, Juncal Formation.
Provincial Range: Late Paleocene ("Meganos Stage") to late Eocene ("Tejon Stage").

Dentalium stramineum Gabb

Dentalium stramineum Gabb, 1864:139, 140, pl. 21, fig. 101. Anderson and Hanna, 1925:144, 145, text fig. 6.
Dentalium (*Dentalium* ?) *stramineum* Gabb. Vokes, 1939:104.
Hypotype: UCR 4679/28, locality 4679.
Local occurrence: *Turritella uvasana applinae* and *Ectinochilus canalifer* faunas, Juncal and Matilija formations.
Provincial Range: Late Paleocene ("Meganos Stage") to late Eocene ("Tejon Stage").

Dentalium cf. *D. calafium* Vokes

Hypotype: UCR 4753/151, locality 4753, *Ectinochilus supraplicatus* fauna, Juncal Formation.

Two specimens that lack longitudinal ribbing and appear to represent Vokes's (1939:105, pl. 16, figs. 30, 31) species were collected from locality 4753.

Class GASTROPODA
Superfamily TROCHACEA
Family TROCHIDAE
Subfamily SOLARIELLINAE
Genus *Solariella* Wood, 1842

Type species: (by original designation) *Solariella maculata* Wood.

Solariella crenulata (Gabb)

(Pl. 5, figs. 1, 2)

Margaritella crenulata Gabb, 1864:118, 119, pl. 20, fig. 74.
Solariella? creunlata (Gabb). Stewart, 1926:317, pl. 26, figs. 10, 10*a*.
Solariella crenulata (Gabb). Hanna, 1927:300, pl. 45, figs. 4, 6, 8.
Hypotype: UCR 4681/201, locality 4681.
Local occurrence: *T. uvasana applinae* fauna, Juncal Formation.
Provincial Range: Middle Eocene ("Domengine Stage").

Family Turbinidae
Genus *Homalopoma* Carpenter, 1864

Type species: (by monotypy) *Turbo sanguineus* Linné.

Homalopoma umpquaensis Merriam and Turner s.s.
(Pl. 5, fig. 3)

Homalopoma umpquaensis Merriam and Turner, 1937:104, pl. 6, fig. 6. Turner, 1938:95, 96, pl. 15, fig. 14.

Hypotypes: UCR 4675/4, locality 4675; UCR 4676/6, locality 4676; UCR 4679/23, locality 4679.

Local occurrence: T. uvasana applinae fauna, Juncal Formation.

Provincial Range: Early Eocene ("Capay Stage") to middle Eocene ("Domengine Stage").

Superfamily NERITACEA
Family Neritidae
Subfamily Neritinae
Genus *Nerita* Linné, 1758

Type species: (by subsequent designation, Montfort, 1810) *Nerita peloronta* Linné.

Subgenus *Theliostyla* Mörch, 1852

Type species: (by subsequent designation, Kobelt, 1879) *Nerita albicilla* Linné.

Nerita (Theliostyla) triangulata Gabb
(Pl. 5, fig. 4)

Nerita (Theliostyla) triangulata Gabb, 1869:170, pl. 28, figs. 52, 52*a*. Vokes, 1939:182, pl. 22, figs. 31, 33, 34.

Nerita triangulata Gabb. Arnold, 1909:14, pl. 4, fig. 12. Hanna, 1927:301, pl. 46, figs. 11, 12, 16, 17.

Hypotype: UCR 4747/3, locality 4747.

Local occurrence: Ectinochilus canalifer fauna, Matilija Sandstone.

Provincial Range: Middle Eocene ("Domengine Stage") to late Eocene ("Tejon Stage").

Genus *Velates* de Montfort, 1810

Type species: (by original designation) *Velates conoideus* de Montfort, 1810 (= *Nerita perversa* Gmelin, 1791).

Velates perversus (Gmelin)
(Pl. 5, figs. 5, 6, 13)

Nerita perversa Gmelin, 1789, *in* Linné, Syst. Nat., 13th ed., vol. 1, pt. 6, p. 3686.

Velates conoideus de Montfort, 1810, Conch. Syst., 2:355.

Velates perversus (Gmelin). Vokes, 1935*b*:382, 383, pl. 25, figs. 1–5; pl. 26, figs. 1, 2.

Hypotypes: UCR 4668/31, 4668/33, 4668/34, locality 4668.

Local occurrence: Turritella uvasana infera fauna, Juncal Formation.

Provincial Range: Early Eocene ("Capay Stage") to middle Eocene ("Domengine Stage").

Numerous specimens of this distinctive species were collected from locality 4668.

Superfamily CERITHIACEA
Family Turritellidae
Genus *Turritella* Lamarck, 1799

Type species: (by monotypy) *Turbo terebra* Linné.

Subgenus? (*Turritella andersoni* stock of Merriam)
Turritella andersoni Dickerson s.s.
(Pl. 5, figs. 7–10)

Turritella andersoni Dickerson, 1916:501, 502, pl. 42, figs. 9*a*, 9*b*. Turner, 1938:83, pl. 22, figs. 4–6. Vokes, 1939:160, 161. Merriam, 1941:76, 77, pl. 9, figs. 1, 2; pl. 10, figs. 1, 3–5, 8; pl. 12, figs. 1–3.

Hypotypes: UCR 4659/201, 4659/202, 4659/203, locality 4659; UCR 4667/201, locality 4667.
Local occurrence: Turritella uvasana infera fauna, Juncal Formation.
Provincial Range: Early Eocene ("Capay Stage").

Typical *Turritella andersoni* is characterized by a whorl profile that is broadly and shallowly concave medially between a pair of broadly spaced weak primary spiral ribs. The whorl surface below the anterior primary is a distinctly flattened basal band and is also somewhat flattened between the posterior primary and the suture. A third weak primary spiral rib occurs intermediate between the anterior and posterior primaries, a secondary thread is usually present between the posterior primary and the suture, and as many as four secondary threads may be present on the concave median area (generally on the posterior half) and may approach weak primary strength.

Specimens of *Turritella andersoni* s.s. in the Pine Mountain section agree closely with Merriam's (1941:76) description of this subspecies but attain a somewhat larger size, averaging about 55 mm in length with a body whorl diameter of about 10 mm. At two localities (UCR 4664 and 4666), a few very large specimens were obtained. These specimens range from 85 to 95 mm in length and have a body whorl diameter of 14 mm. The later whorls of these specimens show slight overhang and separation at the suture, approaching *Turritella andersoni lawsoni* in appearance. Another feature of *T. andersoni* s.s. in the Pine Mountain section is the presence of small nodes on the primary spirals. Although not mentioned by Merriam (1941), similar nodes, somewhat weakly developed, also occur on specimens of this subspecies at its type locality. Nodose spiral ornamentation is present in many members of the *Turritella andersoni* stock and appears to be a persistent and characteristic feature of this group.

The typical subspecies is very abundant in the *Turritella uvasana infera* fauna in the Pine Mountain section. It is restricted to the "Capay Stage" and appears to be directly ancestral to *T. andersoni lawsoni* of the "Domengine Stage." These two subspecies occur in stratigraphic sequence in the Pine Mountain section and also in the Eocene section near Coalinga, California (Vokes, 1939).

Turritella andersoni lawsoni Dickerson
(Pl. 5, figs. 11, 12, 14)

Turritella lawsoni Dickerson, 1916:502, pl. 42, figs. 10*a*, 10*b*. Hanna, 1927:308, pl. 49, fig. 5.
Turritella andersoni lawsoni Dickerson. Vokes, 1939:161. Merriam, 1941:77, 78, pl. 9, figs. 3–8; pl. 12, fig. 4.

Hypotypes: UCR 4694/10, locality 4694; UCR 4700/10, locality 4700; UCR 4701/11, locality 4701.
Local occurrence: Turritella uvasana applinae fauna, Juncal Formation.
Provincial Range: Middle Eocene ("Domengine Stage").

Numerous specimens of this subspecies were collected from several localities in the *Turritella uvasana applinae* fauna. It is distinguished from its apparent ancestor, *T. andersoni* s.s., by its generally much larger size, well-rounded basal quarter of the whorl, and pronounced tendency toward sutural separation and overhang of the adult whorls. There is also a tendency for the spiral ornamentation to become obsolete and the growth lines very pronounced on the adult whorls. Some specimens of *T. andersoni lawsoni* from the Pine Mountain section are characterized by the presence of many subequal nodose spiral ribs on the adult whorls (see pl. 6, fig. 12) and represent the form described by Merriam (1941:78, pl. 9, fig. 9) as variety *secondaria*. At UCR locality 4700, only this variety is present. At the other localities, it is associated with specimens referable to *T. andersoni lawsoni* s.s.

Turritella schencki Merriam s.s.

Turritella schencki Merriam, 1941:81, pl. 10, figs. 9, 10.

Hypotype: UCR 4743/1, locality 4743.

Local occurrence: Ectinochilus canalifer fauna, Matilija Sandstone.

Provincial Range: Late Eocene ("Tejon Stage").

A single well-preserved whorl, showing the characteristic ornamentation of this species, was collected from locality 4743.

Turritella scrippsensis Hanna

(Pl. 6, fig. 1)

Turritella scrippsensis Hanna, 1927:308, pl. 49, figs. 6, 10. Merriam, 1941:81, pl. 9, figs. 15, 16. Kelley, 1943:9, pl. 1, fig. 10. Kleinpell and Weaver, 1963:184, pl. 23, fig. 7.

Hypotypes: UCR 4690/10 and 4690/11, locality 4690.

Local occurrence: Turritella uvasana applinae fauna, Juncal Formation.

Provincial Range: Middle Eocene ("Domengine" and "Transition" "Stages").

This species resembles *Turritella andersoni lawsoni* variety *secondaria* Merriam in having numerous subequal nodose spiral threads on the adult whorls, but is distinguished from that form and from typical *T. andersoni lawsoni* by its broader pleural angle and flat to very shallowly concave whorl profile. It was collected only from locality 4690 in the Pine Mountain section.

Subgenus? (*Turritella buwaldana* stock of Merriam)

Turritella buwaldana Dickerson s.s.

(Pl. 5, fig. 15)

Turritella buwaldana Dickerson, 1916:500, 501, pl. 42, figs. *7a*, *7b*. Hanna, 1927:307, pl. 49, figs. 7, 8, 12. Merriam, 1941: 86, 87, pl. 21, figs. 3–9; pl. 22, figs. 1–14.

Hypotypes: UCR 4702/2, locality 4702; UCR 4705/201, locality 4705; UCR 4743/10, locality 4743.

Local occurrence: Turritella uvasana applinae, Ectinochilus supraplicatus, and *E. canalifer* faunas, Juncal and Matilija formations.

Provincial Range: Middle Eocene ("Domengine Stage") to late Eocene ("Tejon Stage").

The typical subspecies is characterized by four to six principal spiral ribs (consisting of three primaries and one to three posterior secondaries) on the adult whorls with weaker tertiary ribs on the interspaces. On some specimens, particularly the larger forms, all the ribs are of approximately equal strength on the adult whorls. This subspecies is abundant in the *Turritella uvasana applinae* and *Ectinochilus supraplicatus* faunas in the Pine Mountain section. A single specimen was collected from the *E. canalifer* fauna.

Turritella buwaldana crooki Merriam and Turner

(Pl. 5, figs. 16, 17)

Turritella buwaldana crooki Merriam and Turner, 1937:105, pl. 5, fig. 6. Merriam, 1941:87, pl. 21, figs. 1, 2.

Hypotypes: UCR 4656/201 and 4656/202, locality 4656.

Local occurrence: Turritella uvasana infera fauna, Juncal Formation.

Provincial Range: Early Eocene ("Capay Stage").

This subspecies is fairly common in the *Turritella uvasana infera* fauna in the Pine Mountain section. It is distinguished from the typical subspecies by the character of the spiral ribbing. In *T. buwaldana crooki,* the adult ornamentation consists of three widely spaced primary spirals on the anterior two-thirds of the whorl, followed by three closely spaced secondaries on the posterior third. The primary spirals are generally more strongly developed than the secondary spirals, although the first posterior secondary may approach the primaries in strength. Weak tertiary interribs occur between the primary and secondary spirals and occasionally approach them in strength, although this tendency is less pronounced than in the typical subspecies.

The stratigraphic distribution of *Turritella buwaldana* s.s. and *T. buwaldana crooki* in the Pine Mountain section and elsewhere on the Pacific Coast suggests that these taxa may form an evolutionary sequence. *Turritella buwaldana crooki* has been reported only from the "Capay Stage"; *T. buwaldana* s.s. has its lowest-known occurrence in the "Domengine Stage."

Subgenus? (*Turritella uvasana* stock of Merriam)

Turritella uvasana Conrad s.s.

(Pl. 6, fig. 2)

Turritella uvasana Conrad, 1855:10; 1857:321, pl. 2, fig. 12. Merriam, 1941:88, 89, pl. 15, figs. 1, 2; pl. 16, fig. 15.

Hypotypes: UCR 4716/10, locality 4716; UCR 4718/10, locality 4718; UCR 4722/210, locality 4722.

Local occurrence: Ectinochilus canalifer fauna, Juncal and Matilija formations.

Provincial Range: Late Eocene ("Tejon Stage").

Turritella uvasana is represented in the Pine Mountain section by five subspecies, including the typical. These five subspecies occur in stratigraphic sequence and appear to form an evolutionary sequence. Elsewhere on the Pacific Coast, wherever two or more of these subspecies have been reported from the same section, they occur in the same sequence as in the Pine Mountain section. As would be expected of members of an evolutionary sequence, these subspecies intergrade to some extent but each is distinguished by a particular set of morphologic characteristics.

According to Merriam (1941:42), members of the *Turritella uvasana* stock are characterized by a pair of closely spaced primary spirals on the nuclear whorls, an evenly convex whorl profile (although some members are characterized by a flat profile), and a growth line with a deep antispiral sinus (except in *Turritella variata*). The ornamentation in the adult consists of about six spiral threads of approximately equal strength, or with the two primaries on the lower part of the whorl somewhat stronger. Fine interribs may or may not be present between the principal spirals.

The typical subspecies of *Turritella uvasana* is distinguished by the persistence of the pair of primary spirals as the most prominent elements of sculpture on the shell. With growth, a secondary spiral between the primary pair approaches them in strength, giving the adult whorls a characteristic tricarinate appearance. The secondary spirals posterior to the primary pair are of moderate strength and decrease progressively in size toward the posterior suture. Fine tertiary threads frequently occur on the interspaces between the primaries and secondaries. The whorl profile varies from evenly convex to slightly constricted above. *Turritella uvasana* s.s. is a relatively large form. Specimens from the Pine Mountain section average about 70 mm in length, with an average body whorl diameter of about 15 mm.

The typical subspecies is confined to the lower part of the *Ectinochilus canalifer* fauna in the Pine Mountain section. It intergrades with and appears to have evolved from *Turritella uvasana neopleura*. It is replaced in the upper part of the *Ectinochilus canalifer* fauna by its probable direct descendant, *Turritella uvasana sargeanti*.

Turritella uvasana infera Merriam

(Pl. 6, figs. 5–7)

Turritella uvasana infera Merriam, 1941:90, pl. 40, figs. 2–4.

?*Turritella uvasana hendoni* Merriam var. B Turner, 1938:85, pl. 22, figs. 7, 10, 12, 13.

?*Turritella uvasana* Conrad n. var. Crowell and Susuki, 1959:588, pl. 2, fig. 10.

Hypotypes: UCR 4659/103, locality 4659; UCR 4661/41, locality 4661; UCR 4662/201, locality 4662.

Local occurrence: Turritella uvasana infera fauna, Juncal Formation.

Provincial Range: Late Paleocene ("Meganos Stage") to early Eocene ("Capay Stage").

This subspecies is very abundant in the *Turritella uvasana infera* fauna. It is the earliest known member of the *Turritella uvasana* stock and, therefore, is probably ancestral to all other members of this group. It is characterized by its small size (specimens from the Pine Mountain section average about 45 mm in length and have a body whorl diameter of about 10 mm), flat to gently convex whorl profile, and adult whorls that are generally ornamented by no more than six (and often only five) principal spiral ribs of approximately equal strength. The two primary spirals are more closely spaced and are situated nearer the anterior suture than in the other subspecies of *T. uvasana*. Tertiary interribs are only weakly developed on the interspaces between the principal spirals and are often absent. Occasional specimens of *T. uvasana infera* have a moderately convex whorl profile and approach *T. uvasana applinae* in appearance.

At the present time, *Turritella uvasana infera* has been identified with certainty at only three localities: the type locality in the basal conglomerate member of the Llajas Formation in the Simi Valley of Ventura County; the upper part of the underlying Santa Susana Formation; and the *Turritella uvasana infera* fauna in the Pine Mountain section. It may also be represented, however, by *Turritella uvasana* n. var. Crowell and Susuki (1959) from the lower

Maniobra Formation in eastern Riverside County and by *T. uvasana hendoni* var. B Turner (1938) from the lower Umpqua Formation in southern Oregon. Both of these forms resemble *T. uvasana infera* in their small size, in having no more than six principal spirals on the adult whorls, and in the close spacing and relatively anterior position of the primary spirals. They are distinguished from *T. uvasana infera* only by a slightly more convex whorl profile.

Turritella uvasana applinae Hanna

(Pl. 6, figs. 3, 4; pl. 7, fig. 19)

Turritella applini Hanna, 1927:307, pl. 49, figs. 1, 4.
Turritella uvasana applini Hanna. Merriam, 1941:93, 94, pl. 16, figs. 5, 6; pl. 18, fig. 2.
Turritella uvasana applinae Hanna. Keen and Bentson, 1944:210.
?*Turritella uvasana hendoni* Merriam var. A Turner, 1938:84, pl. 22, figs. 11, 14.

Hypotypes: UCR 4688/101, locality 4688; UCR 4694/21, locality 4694; UCR 4702/101, locality 4702.

Local occurrence: Turritella uvasana applinae fauna, Juncal Formation.

Provincial Range: Middle Eocene ("Domengine Stage").

This subspecies is abundant in the *Turritella uvasana applinae* fauna in the Pine Mountain section. It is characterized by its small size (rarely exceeding 50 mm in length), evenly convex whorl profile, and delicate sculpture. The ornamentation in the adult consists of six to nine principal spiral ribs, with the two primaries, an intermediate secondary and the first three posterior secondaries being of approximately equal strength and spacing. In some specimens, the primaries and the intermediate secondary are somewhat stronger than the other spirals, as in the typical subspecies of *T. uvasana.* Fine tertiary threads may occur intermittently on the interspaces between the principal spirals or may be lacking entirely.

Turritella uvasana applinae is distinguished from its probable ancestor, *T. uvasana infera,* by a more convex whorl profile and a greater number of principal spirals on the adult whorls. From its apparent descendant, *T. uvasana neopleura,* it is distinguished by smaller size, more delicate sculpture, and fewer and more prominent spiral ribs.

Turritella uvasana hendoni var. A Turner (1938: 84) from the Tyee Formation in southern Oregon closely resembles *T. uvasana applinae* and may be synonymous with it.

Turritella uvasana neopleura Merriam

(Pl. 6, figs. 8–11)

Turritella uvasana Conrad var. *neopleura* Merriam, 1941:93, pl. 15, figs. 4–7.

Hypotypes: UCR 4706/201 and 4706/204, locality 4706; UCR 4743/2, locality 4743; UCR 4744/21 and 4744/22, locality 4744.

Local occurrence: Ectinochilus supraplicatus and *E. canalifer* faunas, Juncal and Matilija formations.

Provincial Range: Late middle Eocene ("Transition Stage") to late Eocene ("Tejon Stage").

This subspecies is characterized by its large size (often exceeding that of the typical subspecies), gently and regularly convex to nearly flat whorl profile, and sculpture of numerous relatively fine subequal spiral threads on the adult whorls. The primary spirals are generally not distinguishable as such beyond the adolescent stage. The separation at the suture is generally obvious, exposing the base of the whorl above. A basal groove is often well developed below the primaries in the early growth stages, but may become almost obsolete on the latest adult whorls. A well-defined sutural rib is often present at the base of the whorl.

Turritella uvasana neopleura is locally abundant in the lower part of the *Ectinochilus canalifer* fauna. Preservation of the specimens, however, is generally poor. A few specimens were obtained from the *Ectinochilus supraplicatus* fauna.

Merriam (1941:93) considered this form to be only a variety of typical *Turritella uvasana.* The evidence from the Pine Mountain section, however, indicates that *T. uvasana neopleura* occupies a stratigraphic interval mainly below that characterized by the typical subspecies and is apparently ancestral to that form. It is therefore considered herein to be a distinct chronological subspecies within the *T. uvasana* lineage.

Turritella uvasana sargeanti Anderson and Hanna

(Pl. 6, figs. 12–16)

Turritella sargeanti Anderson and Hanna, 1925:125.

Turritella uvasana sargeanti Anderson and Hanna. Merriam, 1941:96, 97, pl. 16, figs. 1–3.

Hypotypes: UCR 4726/101, 4726/102, 4726/103, locality 4726; UCR 4732/301, 4732/302, locality 4732.

Local occurrence: Ectinochilus canalifer fauna, Matilija Sandstone.

Provincial Range: Late Eocene ("Tejon Stage").

This distinctive subspecies is characterized by its large size and by the strong development of the two primary spirals, the upper of which is frequently the median carina or heaviest spiral of the adult. The presence of smooth, fairly broad interspaces between the principal spirals is also a characteristic feature. Tertiary interribs are weakly developed or lacking. The early whorls of this subspecies increase more rapidly in size than in the other subspecies of *T. uvasana,* resulting in a broader apical angle.

Turritella uvasana sargeanti occurs only in the upper half of the *Ectinochilus canalifer* fauna in the Pine Mountain section and appears to be a direct descendant of *T. uvasana* s.s. These two subspecies also occur in stratigraphic sequence in the type Tejon Formation (Merriam, 1941:97).

Turritella variata sanmarcosensis Kleinpell and Weaver (1963:185–186, pl. 24, figs. 2–4) from the *Turritella schencki delaguerrae* Zone in the Santa Ynez range in western Santa Barbara County, California, is very closely related to *Turritella uvasana sargeanti* and appears to have evolved from it. *T. variata sanmarcosensis,* according to its authors, is distinguished from *T. uvasana sargeanti* by the equal development of the primary spirals (as noted above, in typical *T. uvasana sargeanti* the posterior primary is more strongly developed, producing a medially carinate whorl profile). Specimens of *T. uvasana sargeanti* from the upper end of its local range-zone in the Pine Mountain section (see pl. 7, figs. 15, 16) resemble *T. variata sanmarcosensis* in having equally developed primary spirals but differ in having four, rather than three, secondary spirals posterior to the primary pair. The number of secondary spirals posterior to the primaries appears to provide a useful basis for distinguishing between *Turritella uvasana* s.l. and *T. variata* s.l. All subspecies of *T. uvasana* have at least four posterior secondary spirals, whereas *T. variata* and its subspecies have no more than three.

Family Architectonicidae

Genus *Architectonica* Röding, 1798

Type species: (by subsequent designation, Gray, 1847) ***Trochus perspectivus*** **Linné.**

Subgenus *Architectonica* s.s.

Architectonica (*Architectonica*) *hornii* Gabb

(Pl. 7, figs. 5–7)

Architectonica hornii **Gabb, 1864:117, 118, pl. 29, figs. 224, *a*, *b*. Anderson and Hanna, 1925: 123, 124, pl. 8, fig. 2; pl. 9, fig. 1. Stewart, 1926:343, pl. 30, fig. 13.**

Hypotype: UCR 4741/60, locality 4741.

Local occurrence: Ectinochilus canalifer fauna, Juncal and Matilija formations.

Provincial Range: Late Eocene ("Tejon Stage").

This species is characterized by a double keel on the periphery of the whorls (with the upper of the two carinae on the keel being the more prominent), by small crenulations on the whorls just below the suture, and by two prominent crenulated spiral ridges around the margin of the umbilicus. Except for the small crenulations and a spiral line of variable strength just above the keel, the dorsal surface of the whorls is ornamented only by fine growth lines.

Subgenus *Stellaxis* Dall, 1892

Type species: (by original designation) *Solarium alveatum* Conrad.

Architectonica (*Stellaxis*) *cognata* Gabb

(Pl. 7, figs. 1–3)

Architectonica cognata Gabb, 1864, p. 117, pl. 20, figs. 72, 72*a*, *c* [not 72*b*, = *A.* (*S.*) *alveata* (Conrad)].

Stellaxis cognata (Conrad). Waring, 1917: 98.

Architectonica (*Stellaxis*) *cognata* Gabb. Stewart, 1926:343, pl. 28, figs. 7, 8; 1946, pl. 11, fig. 4. Turner, 1938:90, pl. 18, fig. 17. Vokes, 1939:163, 164.

Hypotype: UCR 4750/20, locality 4750.

Local occurrence: Turritella uvasana applinae fauna, Juncal Formation.

Provincial Range: Early Eocene ("Capay Stage") to middle Eocene ("Domengine Stage").

This species is distinguished from *A. hornii* by its larger size, single keel on the periphery of the whorls, and absence of small crenulations on the whorls below the suture. Two specimens were collected, one from UCR locality 4672 and the other from UCR locality 4750.

Subgenus *Solariaxis* Dall, 1892

Type species: (by original designation) *Solarium elaboratum* Conrad.

Architectonica (*Solariaxis*) *ullreyana* Dickerson

(Pl. 7, fig. 9)

Architectonica ullreyana Dickerson, 1916:487, 488, pl. 40, figs. 5*a*, 5*b*.

Hypotypes: UCR 4688/18, locality 4688; UCR 4690/9, locality 4690.

Local occurrence: Turritella uvasana applinae fauna, Juncal Formation.

Provincial Range: Early Eocene ("Capay Stage") to middle Eocene ("Domengine Stage").

This distinctive species is characterized by the presence of numerous crenulated spiral ribs on the dorsal and ventral surfaces of the whorls and by a very acute keel. The ventral surface of the body whorl adjacent to the keel is slightly excavated. Two specimens, both encrusted with a hard limestone matrix, were collected from the Pine Mountain section. The species was described from the type locality of Dickerson's (1913; 1916) *Siphonalia sutterensis* Zone at Marysville Buttes. It also occurs in the Llajas Formation in the Simi Valley (Givens: unpub. data).

Architectonica ullreyana appears to be closely related to *A.* (*Solariaxis*) *acuta* Conrad from the Claibornian (middle Eocene) and Jacksonian (late Eocene) Stages of the Gulf Coast (see Palmer, 1937:167, pl. 20, figs. 14–17; Harris and Palmer, 1946:275, 276, pl. 33, figs. 9–11, 14–19; pl. 65, figs. 1, 2).

Family Cerithiidae

Genus *Cerithium* Bruguiere, 1789

Type species: (by original designation) *Cerithium nodulosum* Bruguiere.

Cerithium cliffensis Hanna

(Pl. 7, fig. 4)

Cerithium cliffensis Hanna, 1927:310, pl. 50, figs. 1–4.

Hypotypes: UCR 4675/81, locality 4675; UCR 4703/30, locality 4703; UCR 4706/7, locality 4706.

Local occurrence: Turritella uvasana applinae and *Ectinochilus supraplicatus* faunas, Juncal Formation.

Provincial Range: Middle Eocene ('Domengine" and "Transition" "Stages").

Cerithium excelsum (Dall)

Cerithiopsis excelsus Dall, 1909:75, pl. 3, fig. 9.

Cerithiopsis excelsa Dall. Bartsch, 1911:352, pl. 36, fig. 1. Turner, 1938:82, pl. 21, fig. 6.

Hypotypes: UCR 4667/90, locality 4667; UCR 4668/7, locality 4668.

Local occurrence: Turritella uvasana infera fauna, Juncal Formation.

Provincial Range: Early Eocene ("Capay Stage") to middle Eocene ("Domengine Stage").

Cerithium cf. *C. orovillensis* (Dickerson)

Hypotype: UCR 4667/6, locality 4667, *Turritella uvasana infera* fauna, Juncal Formation.

Two poorly preserved specimens comparable with Dickerson's (1916:489, 490, pl. 39, fig. 7) species were collected from locality 4667.

Genus *Campanilopa* Iredale, 1917

Type species: (by original designation) *Cerithium giganteum* Lamarck.

Campanilopa dilloni Hanna and Hertlein

(Pl. 7, fig. 10)

Campanilopa dilloni Hanna and Hertlein, 1949:393, pl. 77, figs. 2, 4; text fig. 1.

Hypotype: UCR 4668/101, locality 4668.

Local occurrence: Turritella uvasana infera fauna, Juncal Formation.

Provincial Range: Early Eocene ("Capay Stage").

Two broken specimens of this species were collected from locality 4668.

Family Potamididae

Subfamily Potamidinae

Genus *Potamides* Brongniart, 1810

Type species: (by monotypy) *Potamides lamarcki* Brongniart.

Potamides carbonicola Cooper

Potamides carbonicola **Cooper, 1894:44, pl. 1, figs. 14–19 [listed as *Cerithidea carbonicola* on plate legend]. Arnold, 1909:14, pl. 4, figs. 2, 3. Arnold and Anderson, 1910:71, pl. 26, figs. 2, 3. Hanna, 1927:312, pl. 55, figs. 2, 7, 9. Turner, 1938:82, pl. 21, fig. 1. Vokes, 1939:157, 158, pl. 20, figs. 23–27.**

Hypotype: UCR 4714/81, locality 4714.

Local occurrence: *Ectinochilus supraplicatus* and *E. canalifer* faunas, Juncal and Matilija formations.

Provincial Range: Early Eocene ("Capay Stage") to late Eocene ("Tejon Stage").

Family THIARIDAE
Genus *Loxotrema* Gabb, 1868

Type species: (by monotypy) *Loxotrema turrita* Gabb.

Loxotrema turritum Gabb
(Pl. 6, fig. 17)

Loxotrema turrita Gabb, 1868:147, pl. 14, fig. 21; 1869:168, pl. 28, fig. 49. Arnold, 1909:14, pl. 4, fig. 17. Hanna, 1927:312, pl. 50, figs. 5–8. Vokes, 1939:159, pl. 20, figs. 15–19.

Loxotrema turritum Gabb. Stewart, 1926:347, 348, pl. 26, fig. 3, 4. Turner, 1938:81, pl. 17, figs. 12, 13. Keen and Bentson, 1944:168.

Pachychilus (?*Loxotrema*) *turritum* (Gabb). Wenz, 1939:686.

Hypotype: UCR 4747/401, locality 4747.

Local occurrence: *Ectinochilus supraplicatus* and *E. canalifer* faunas, Juncal and Matilija formations.

Provincial Range: Early Eocene ("Capay Stage") to late Eocene ("Tejon Stage").

Family THIARIDAE?
Genus uncertain
"Trichotropis" lajollaensis Hanna
(Pl. 6, fig. 18)

Trichotropis? lajollaensis Hanna, 1927:311, 312, pl. 48, figs. 4–6, 9, 11.

Hypotype: UCR 4747/501, locality 4747.

Local occurrence: *Ectinochilus canalifer* fauna, Matilija Sandstone.

Provincial Range: Middle Eocene ("Domengine Stage") to late Eocene ("Tejon Stage").

The genus and family of this species are uncertain. It lacks the deep umbilical perforation characteristic of *Trichotropis*. The character of the aperture and spire suggest a close relationship with the genus *Thiara* Bolten, *in* Röding, 1798.

Superfamily SCALACEA
Family SCALIDAE
Genus *Scalina* Conrad, 1865

Type species: (by subsequent designation, Palmer, 1937) *Scala* (*Scalina*) *staminea* Conrad.

Scalina cf. *S. aragoensis* (Durham)

Hypotype: UCR 4705/12, locality 4705, *Ectinochilus supraplicatus* fauna, Juncal Formation.

Two poorly preserved specimens comparable with Durham's (1937:506, 507, pl. 57, fig. 24) species were collected from locality 4705.

Superfamily CALYPTRACEA
Family CALYPTRAEIDAE
Genus *Calyptraea* Lamarck, 1799

Type species: (by monotypy) *Patella chinensis* Linné.

Calyptraea diegoana (Conrad)

Trochita diegoana Conrad, 1855:7, 17; 1857:319, 327, pl. 5, fig. 42.
Galerus excentricus Gabb, 1864:136, pl. 20, fig. 95; pl. 29, fig. 232*a*.
Calyptraea diegoana (Conrad). Stewart, 1926:340, 341, pl. 27, fig. 15. Turner, 1938:89, 90, pl. 20, figs. 1, 2. Weaver, 1942:351, 352, pl. 71, figs. 16, 20; pl. 103, fig. 3. Kleinpell and Weaver, 1963:186, pl. 24, fig. 7. Hickman, 1969:79, pl. 11, figs. 7, 8.

Hypotypes: UCR 4675/13, locality 4675; UCR 4706/11, locality 4706; UCR 4720/10, locality 4720.

Local occurrence: Turritella uvasana infera, T. uvasana applinae, Ectinochilus supraplicatus, and *E. canalifer* faunas, Juncal and Matilija formations.

Provincial Range: Late Paleocene ("Meganos Stage") to early Oligocene (Eugene Formation of Oregon).

Genus *Crepidula* Lamarck, 1799

Type species: (by monotypy) *Patella fornicata* Linné.

Subgenus *Spirocrypta* Gabb, 1864

Type species: (by monotypy) *Crypta (Spirocrypta) pileum* Gabb.

Crepidula (Spirocrypta) pileum (Gabb)

Crypta (Spirocrypta) pileum Gabb, 1864:137, pl. 29, figs. 233, 233*a*, *b*.
Crepidula pileum (Gabb). Anderson and Hanna, 1925:122, pl. 13, fig. 7. Stewart, 1926:341, 342, pl. 29, figs. 2, 3. Kleinpell and Weaver, 1963:186, 187, pl. 24, figs. 8, 10, 11.
Crepidula (Spirocrypta) pileum (Gabb). Clark, 1938:701, pl. 4, fig. 19.

Hypotypes: UCR 4717/9, locality 4717; UCR 4719/7, locality 4719.

Local occurrence: Ectinochilus canalifer fauna, Juncal and Matilija formations.

Provincial Range: Late Eocene ("Tejon Stage" to *Turritella variata lorenzana* Zone of Kleinpell and Weaver, 1963).

Family XENOPHORIDAE
Genus *Xenophora* Fischer von Waldheim, 1807

Type species: (by subsequent designation, Gray, 1847) *Trochus conchyliophorus* Born.

Xenophora stocki Dickerson

(Pl. 7, fig. 8)

Xenophora stocki Dickerson, 1916:502, 503, pl. 37, figs. 4*a*, 4*b*. Hanna, 1927:306. Stewart, 1946, table 1.

Hypotype: UCR 4680/7, locality 4680.

Local occurrence: T. uvasana applinae fauna, Juncal Formation.

Provincial Range: Middle Eocene ("Domengine Stage").

Superfamily STROMBACEA
Family STROMBIDAE
Genus *Ectinochilus* Cossmann, 1889

Type species: (by original designation) *Strombus canalis* Lamarck.

Subgenus *Macilentos* Clark and Palmer, 1923

Type species: (by original designation) *Rimella macilenta* White, 1889.

Ectinochilus (Macilentos) macilentus (White)

(Pl. 7, figs. 13, 16)

Rimella macilenta White, 1889:19, pl. 3, figs. 10–12.
Ectinochilus (Macilentos) macilentus (White). Clark and Palmer, 1923:280, pl. 51, figs. 9, 10.
Rimella (Macilentos) macilenta White. Vokes, 1939:155, 156, pl. 20, figs. 1, 2, 4, 5.
Ectinochilus macilentus (White). Stewart, 1946:93, pl. 11, figs. 12–15.

Hypotypes: UCR 4679/7, locality 4679; UCR 4680/17, locality 4680; UCR 4655/7, locality 4655; UCR 4666/6, locality 4666.

Local occurrence: Turritella uvasana infera and *T. uvasana applinae* faunas, Juncal Formation.

Provincial Range: Early Eocene ("Capay Stage") to middle Eocene ("Domengine Stage").

Subgenus *Cowlitzia* Clark and Palmer, 1923

Type species: (by original designation) *Cowlitzia washingtonensis* Clark and Palmer.

Ectinochilus (Cowlitzia) canalifer (Gabb)

(Pl. 7, fig. 11)

Rostellaria canalifer Gabb, 1864:123, 124, pl. 29, fig. 228.
Cowlitzia canalifera (Gabb). Clark and Palmer, 1923:284, pl. 41, figs. 15–20. Anderson and Hanna, 1925:102, 103, pl. 9, figs. 6, 9, 13.
Ectinochilus (Cowlitzia) canalifer (Gabb). Stewart, 1926:366, 367, pl. 29, fig. 8. Kleinpell and Weaver, 1963:189, pl. 25, fig. 6.

Hypotypes: UCR 4714/30, locality 4714; UCR 4715/161, locality 4715.

Local occurrence: E. canalifer fauna, Matilija Sandstone.

Provincial Range: Late Eocene ("Tejon Stage" to *Turritella schencki delaguerrae* Zone of Kleinpell and Weaver, 1963).

Ectinochilus (Cowlitzia) supraplicatus (Gabb)

(Pl. 7, fig. 12)

Rostellaria (Rimella) simplex Gabb, 1864:124, pl. 20, fig. 80.
Cowlitzia simplex (Gabb). Clark and Palmer, 1923:285, pl. 51, figs. 21–24. Hanna, 1927:313, 314, pl. 50, fig. 11.
Ectinochilus canalifer supraplicatus (Gabb). Stewart, 1926:369, 370, pl. 28, fig. 12.

Hypotype: UCR 4753/131, locality 4753.

Local occurrence: Ectinochilus supraplicatus fauna, Juncal Formation.

Provincial Range: Middle Eocene ("Domengine Stage") to late middle Eocene ("Transition Stage").

Several well-preserved specimens of this species were collected from localities 4708 and 4753. It has its highest-known occurrence within the "Transition Stage" and is probably ancestral to *E. canalifer,* from which it is distinguished by its larger size, more inflated shell, and more numerous and closely spaced axial ribs on the spire.

Genus *Tibia* Röding, 1798

Type species: (by subsequent designation, Dall, 1906) *Murex fusus* Linné.

Subgenus *Eotibia* Clark, 1942

Type species: (by original designation) *Rostellaria? lucida* Sowerby.

Tibia (Eotibia) llajasensis Clark

Tibia (Eotibia) llajasensis Clark, 1942:118, pl. 19, figs. 3–6.

Hypotype: UCR 4666/8, locality 4666.

Local occurrence: T. uvasana infera fauna, Juncal Formation.
Provincial Range: Early Eocene ("Capay Stage").

A single specimen of this species was collected from locality 4666.

Genus *Chedevillea* Cossmann, 1906

Type species: (by original designation) *Rimella munieri* Chedeville.

Chedevillea stewarti Clark

Chedevillea stewarti Clark, 1942:117, pl. 19, figs. 7–11.
Hypotypes: UCR 4666/5, locality 4666; UCR 4668/5, locality 4668.
Local occurrence: T. uvasana infera fauna, Juncal Formation.
Provincial Range: Early Eocene ("Capay Stage").

A few specimens of this species were collected from localities 4666 and 4668.

Superfamily NATICACEA
Family NATICIDAE
Subfamily GLOBULARIINAE
Genus *Pachycrommium* Woodring, 1928

Type species: (by original designation) *Amaura guppyi* Gabb.

Pachycrommium? clarki (Stewart)

(Pl. 8, figs. 6, 10)

Amaurellina (*Euspirocrommium*) *clarki* Stewart, 1926:336–339, pl. 26, figs. 8, 9. Turner, 1938:86, pl. 20, fig. 3. Weaver, 1942:345, pl. 70, figs. 10, 18. Kleinpell and Weaver, 1963:188, pl. 27, fig. 15.
Amaurellina clarki Stewart. Gardner and Bowles, 1934:246, figs. 6, 8.
Pachycrommium? clarki (Stewart). Vokes, 1939:175, pl. 22, figs. 11, 30.
Hypotypes: UCR 4674/10, locality 4674; UCR 4699/101, locality 4699; UCR 4707/5, locality 4707; UCR 4715/14, locality 4715.
Local occurrence: Turritella uvasana applinae, Ectinochilus supraplicatus, and *E. canalifer* faunas, Juncal and Matilija formations.
Provincial Range: Early Eocene ("Capay Stage") to late Eocene ("Tejon Stage" plus *Turritella schencki delaguerrae* Zone of Kleinpell and Weaver, 1963).

This species is characterized by a carinate shoulder on the body whorl, with the carina extending back onto the penultimate whorl in some specimens. It resembles *Tejonia moragai,* which also has a carinate shoulder, but is much larger in size and lacks the fine, incised, spiral lines characteristic of *Tejonia.*

Pachycrommium? n. sp.

(Pl. 8, figs. 9, 11)

Hypotype: UCR 4668/11, locality 4668.
Local occurrence: Turritella uvasana infera fauna, Juncal Formation.
Description: shell large, inflated, nonumbilicate; spire high, at least seven whorls, apex broken; shoulder rounded; suture linear, slightly grooved; aperture ovate, biangulate posteriorly, rounded anteriorly, slightly deformed; outer lip moderately thick, broken, inner lip covered by a thin callus; surface of shell smooth except for fine growth lines.
Dimensions of hypotype: length, 56 mm; width, about 35 mm (the body whorl is slightly distorted); length of aperture, 33 mm.

This species is known only from a single specimen. It resembles *P.? clarki,* but lacks the carinate shoulder on the body whorl.

Genus *Tejonia* Hanna and Hertlein, 1943

Type species: (by original designation) *Natica alveata* Conrad.

Tejonia moragai (Stewart)

(Pl. 8, fig. 8)

Natica alveata Conrad, 1855:10, 19; 1857:318, 321, 328, pl. 2, figs. 8, *a*.
Amaurellina moragai Stewart, 1926:334, 335, pl. 28, fig. 3 [new name for *Natica alveata* Conrad, preoccupied]. Kleinpell and Weaver, 1963:188, pl. 25, figs. 1, 2.

Hypotypes: UCR 4706/101, locality 4706; UCR 4723/71, locality 4723.

Local occurrence: Ectinochilus supraplicatus and *E. canalifer* faunas, Juncal and Matilija formations.

Provincial Range: Late middle Eocene ("Transition Stage") to late Eocene ("Tejon Stage" plus *Turritella schencki delaguerrae* Zone of Kleinpell and Weaver, 1963).

Tejonia is closely related to *Amaurellina* "Bayle" Fischer, 1885, but is distinguished by the presence of numerous fine incised spiral lines on the whorls. *T. moragai* is distinguished from all other species of *Tejonia* except *T. malinchae* (Gardner and Bowles, 1934) by the sharply carinate shoulder on the whorls. *T. malinchae,* from the Eocene near Chiapas, Mexico, also has a carinate shoulder but, according to its authors, is somewhat less inflated and rounder in form than *T. moragai.* The relationship of these two species is uncertain.

Tejonia lajollaensis (Stewart)

(Pl. 8, fig. 5)

Amaurellina moragai lajollaensis Stewart, 1926:336, pl. 28, fig. 2.
Tejonia lajollaensis (Stewart). Hanna and Hertlein, 1941:172, fig. 62–30.

Hypotypes: UCR 4675/5, locality 4675; UCR 4703/151, locality 4703.

Local occurrence: Turritella uvasana applinae fauna, Juncal Formation.

Provincial Range: Middle Eocene ("Domengine Stage").

This species is closely related to *Tejonia moragai,* but is distinguished by a rounded to weakly carinate shoulder on the whorls. It is also somewhat larger and more inflated than *T. moragai* and has a broader subsutural shelf. *Tejonia lajollaensis* is confined to stratigraphic horizons below those characterized by *T. moragai* in the Pine Mountain section and elsewhere on the Pacific Coast. It appears to have evolved into that species by the progressive development of a carinate shoulder.

Genus *Ampullella* Cox, 1931

Type species: (by monotypy) *Natica depressa* Lamarck.

Ampullella hewitti Hanna and Hertlein

(Pl. 8, fig. 13)

Ampullella hewitti Hanna and Hertlein, 1949:393, 394, pl. 77, figs. 1, 3; text fig. 2.

Hypotype: UCR 4668/21, locality 4668.

Local occurrence: Turritella uvasana infera fauna, Juncal Formation.

Provincial Range: Early Eocene ("Capay Stage").

This species is abundant at locality 4668. A few specimens were also collected from locality 4662. It has previously been reported only from its type locality in the unnamed sandstone member in the middle of the Lodo Formation near Media Agua Creek in Kern County, California (see Mallory, 1959:76).

Genus *Globularia* Swainson, 1840

Type species: (by subsequent designation, Gray, 1847) *Natica fluctuata* Sowerby.

Subgenus *Eocernina* Gardner and Bowles, 1934

Type species: (by original designation) *Natica hannibali* Dickerson.

Globularia (*Eocernina*) *hannibali* (Dickerson)

(Pl. 9, figs. 1, 3)

Natica hannibali Dickerson, 1914:119, pl. 12, figs. 5*a*, 5*b*; 1916, pl. 38, figs. 9*a*, 9*b*.
Ampullina hannibali (Dickerson). Hanna, 1927:306, pl. 48, figs. 1–3, 10.
Globularia hannibali (Dickerson). Stewart, 1926:331.
Cernina (*Eocernina*) *hannibali* (Dickerson). Turner, 1938:87, 88, pl. 19, fig. 3. Vokes, 1939:172, pl. 22, figs. 1, 3.

Hypotypes: UCR 4658/11, locality 4658; UCR 4699/18, locality 4699; UCR 4707/7, locality 4707; UCR 4708/12, locality 4708.

Local occurrence: Turritella uvasana infera, T. uvasana applinae, and *Ectinochilus supraplicatus* faunas, Juncal Formation.

Provincial Range: Early Eocene ("Capay Stage") to late middle Eocene ("Transition Stage").

Genus *Crommium* Cossmann, 1888

Type species: (by original designation) *Ampullina willemeti* Deshayes.

Crommium andersoni (Dickerson)

(Pl. 8, fig. 3)

Amauropsis andersoni Dickerson, 1914:120, pl. 12, figs. 2*a*, 2*b*.
Amauropsis umpquaensis Dickerson, 1914:120, pl. 12, figs. 3*a*, 3*b*.
Ampullina (*Crommium*) *andersoni* (Dickerson). Turner, 1938:87, pl. 19, figs. 1, 2, 4, 5.
Crommium andersoni (Dickerson). Vokes, 1939:171, pl. 21, figs. 22, 23.

Hypotype: UCR 4659/5, locality 4659.

Local occurrence: Turritella uvasana infera fauna, Juncal Formation.

Provincial Range: Early Eocene ("Capay Stage") to middle Eocene ("Domengine Stage").

This species is abundant at locality 4659. A few specimens were also collected from localities 4662, 4666, and 4667.

Subfamily POLINICINAE

Genus *Polinices* Montfort, 1810

Type species: (by original designation) *Polinices albus* Montfort.

Polinices gesteri (Dickerson)

Natica gesteri Dickerson, 1916:496, pl. 38, fig. 6.
Polinices (*Euspira*) *gesteri* (Dickerson). Clark and Woodford, 1927:120, 121, pl. 21, figs. 14, 15.
Polinices (*Polinices*) *gesteri* (Dickerson). Vokes, 1939:168, pl. 21, figs. 2, 6.

Hypotype: UCR 4674/5, locality 4674.

Local occurrence: Turritella uvasana applinae fauna, Juncal Formation.

Provincial Range: Late Paleocene ("Meganos Stage") to middle Eocene ("Domengine Stage").

A single specimen was collected from locality 4674.

Polinices hornii (Gabb)

(Pl. 8, fig. 2)

Lunatia hornii Gabb, 1864:106, 107, pl. 29, fig. 217. Dickerson, 1915, pl. 4, fig. 11.

Natica hornii **(Gabb). Anderson and Hanna, 1925:115, pl. 10, fig. 7.**
Polinices hornii **(Gabb). Stewart, 1926:324, 325, pl. 30, fig. 15. Turner, 1938:88, pl. 19, figs. 8, 9.**
Polinices (Euspira) hornii **(Gabb). Clark and Woodford 1927:121, pl. 22, figs. 1–4.**

Hypotypes: **UCR 4708/191, locality 4708; UCR 4720/2, locality 4720.**

Local occurrence: Ectinochilus supraplicatus **and *E. canalifer* faunas, Juncal and Matilija formations.**

Provincial Range: Late Paleocene ("Meganos Stage") to late Eocene ("Tejon Stage").

Genus *Neverita* Risso, 1826

Type species: (by monotypy) *Neverita josephina* Risso.

Subgenus *Neverita* s.s.

Neverita (Neverita) globosa Gabb

Neverita globosa Gabb, 1869:161, pl. 27, fig. 39. Dickerson, 1916:510, pl. 39, figs. 5*a*, 5*b*. Stewart, 1926:326, 327, pl. 28, fig. 6. Clark and Woodford, 1927:121, 122, pl. 22, figs. 5–10. Turner, 1938:89, pl. 19, figs. 6, 7, 13–15. Vokes, 1939:169, pl. 21, figs. 9, 15–19.

Hypotype: UCR 4706/291, locality 4706.

Local occurrence: Ectinochilus supraplicatus fauna, Juncal Formation.

Provincial Range: Late Paleocene ("Meganos Stage") to late middle Eocene ("Transition Stage").

Subgenus *Glossaulax* Pilsbry, 1929

Type species: (by original designation) *Neverita reclusianus* (Deshayes).

Neverita (Glossaulax) secta Gabb

Neverita secta Gabb, 1864:108, 109, pl. 29, figs. 220, 220*a*. Dickerson, 1916:510, pl. 39, fig. 6. Stewart, 1926:325, 326, pl. 30, fig. 17.
Natica secta (Gabb). Anderson and Hanna, 1925:117, pl. 10, fig. 9.
Polinices (Neverita) secta (Gabb). Weaver, 1942:341, pl. 70, figs. 3, 4, 7, 8; pl. 100, fig. 30. Kleinpell and Weaver, 1963:188, pl. 24, fig. 15.

Hypotypes: UCR 4714/2 and 4719/14, localities 4714 and 4719.

Local occurrence: Ectinochilus canalifer fauna, Matilija Sandstone.

Provincial Range: Late Eocene ("Tejon Stage" to *Turritella variata lorenzana* Zone of Kleinpell and Weaver, 1963).

Subfamily Sininae

Genus *Sinum* Röding, 1798

Type species: (by subsequent designation, Dall, 1915) *Helix haliotidum* Linné.

Sinum obliquum (Gabb)

Naticina obliqua Gabb, 1864:109, pl. 21, fig. 112. Dickerson, 1915:pl. 5, figs. 5*a*, 5*b*.
Sinum coryliforme Anderson and Hanna, 1925:120, pl. 9, fig. 10; pl. 10, fig. 15; pl. 15, fig. 8.
Sinum obliquum (Gabb). Stewart, 1926:327, pl. 30, fig. 7*a*. Clark, 1938:704, pl. 3, figs. 32, 37. Weaver, 1942:350, 351, pl. 71, fig. 13; pl. 103, fig. 6. Hickman, 1969:85, 86, pl. 11, figs. 9, 10.

Hypotypes: UCR 4655/11, locality 4655; UCR 4675/10, locality 4675; UCR 4707/27, locality 4707; UCR 4720/8, locality 4720.

Local occurrence: Turritella uvasana infera, T. uvasana applinae, Ectinochilus supraplicatus, and *E. canalifer* faunas, Juncal and Matilija formations.

Provincial Range: Early Eocene ("Capay Stage") to early Oligocene (Eugene Formation of Oregon).

Subfamily Naticinae

Genus *Natica* Scopoli, 1777

Type species: (by subsequent designation, Harris, 1897) *Nerita vitellus* Linné.

Subgenus *Natica* s.s.

Natica (*Natica*) *uvasana* Gabb

(Pl. 7, fig. 18; pl. 8, fig. 1)

Natica uvasana Gabb, 1864:212, pl. 32, fig. 277. Anderson and Hanna, 1925:116, pl. 9, figs. 3, 4. Stewart, 1926:322, pl. 30, fig. 14. Kleinpell and Weaver, 1963:187, pl. 24, figs. 12, 13.

Hypotypes: UCR 4706/5, locality 4706; UCR 4714/12, locality 4714; UCR 4715/4, locality 4715.

Local occurrence: Ectinochilus supraplicatus and *E. canalifer* faunas, Juncal and Matilija formations.

Provincial Range: Late middle Eocene ("Transition Stage") to late Eocene ("Tejon Stage").

Subgenus *Carinacca* Marwick, 1924

Type species: (by original designation) *Ampullina waihaoensis* Suter.

Natica (*Carinacca*) *rosensis* Hanna

(Pl. 8, fig. 12)

Natica rosensis Hanna, 1927:305, pl. 47, figs. 7–9.

Natica (*Carinacca*) *rosensis* Hanna. Keen and Bentson, 1944:177.

Hypotype: UCR 4703/7, locality 4703.

Local occurrence: Turritella uvasana applinae fauna, Juncal Formation.

Provincial Range: Middle Eocene ("Domengine Stage").

This species is characterized by a small central funicle and a more prominent basal funicle spiraling into the umbilicus. A few specimens were collected from localities 4675, 4683, 4688, and 4703.

Genus *Euspira* Agassiz, 1839

Type species: (by subsequent designation, Dall, 1908) *Natica glaucinoides* Sowerby (=*Natica labellata* Lamarck).

Euspira clementensis (Hanna)

(Pl. 7, figs. 15, 17)

Natica clementensis Hanna, 1927:304, 305, pl. 47, figs. 1, 3, 4, 6. Vokes, 1939:168.

Polinices (*Euspira*) *clementensis* (Hanna). Clark, 1938:703, pl. 4, figs. 15, 22.

Hypotype: UCR 4688/25, locality 4688.

Local occurrence: Turritella uvasana applinae fauna, Juncal Formation.

Provincial Range: Middle Eocene ("Domengine Stage") to late Eocene ("Tejon Stage").

This species is characterized by a deeply channeled suture and open umbilicus without a funicle. There is a small callus deposit at the posterior end of the inner lip. A few specimens were collected from localities 4685 and 4688.

Euspira nuciformis (Gabb)

(Pl. 7, fig. 14)

Lunatia nuciformis Gabb, 1864:107, pl. 28, fig. 218. Dickerson, 1916:510, pl. 39, fig. 4.

Natica nuciformis (Gabb). Anderson and Hanna, 1925:116, pl. 10, fig. 8.

Euspira nuciformis (Gabb). Stewart, 1926:323, pl. 30, fig. 16.

Polinices (*Euspira*) *nuciformis* (Gabb). Clark and Woodford, 1927:121, pl. 21, figs. 16, 17. Turner, 1938:88, pl. 20, figs. 4, 5. Clark, 1938:703, pl. 4, figs. 26, 31. Vokes, 1939:168, pl. 21, figs. 12–14.

Hypotypes: UCR 4680/15, locality 4680; UCR 4671/2, locality 4671.

Local occurrence: Turritella uvasana applinae fauna, Juncal Formation; *Ectinochilus canalifer* fauna, Matilija and Cozy Dell formations.

Provincial Range: Late Paleocene ("Meganos Stage") to late Eocene ("Tejon Stage").

Superfamily TONNACEA

Family Cassididae

Genus *Galeodea* Link, 1807

Type species: (by original designation) *Buccinum echinophora* Linné.

Subgenus *Gomphopages* Gardner, 1939

Type species: (by original designation) *Galeodea turneri* Gardner.

Galeodea (*Gomphopages*) *sutterensis* Dickerson

(Pl. 8, fig. 4)

Galeodea sutterensis Dickerson, 1916:492, pl. 40, figs. 1*a*, 1*b*. Schenck, 1926:84, pl. 15, figs. 1, 2. Turner, 1938:92, pl. 18, fig. 19. Vokes, 1939:150, 151, pl. 19, fig. 15.

Galeodea (*Gomphopages*) *sutterensis* Dickerson. Durham, 1942:184, pl. 29, fig. 2.

Hypotypes: UCR 4655/4, locality 4655; UCR 4658/5, locality 4658.

Local occurrence: Turritella uvasana infera fauna, Juncal Formation.

Provincial Range: Early Eocene ("Capay Stage").

Two specimens referable to this species were collected from the *Turritella uvasana infera* fauna. *G. sutterensis* is characterized by three nodose carinae on the body whorl, a low spire, and a long anterior canal. The nodes on the posterior carina, which forms the shoulder of the whorl, are spinose.

Galeodea (*Gomphopages*) *susanae* Schenck

Galeodea susanae Schenck, 1926:85, pl. 15, figs. 3–7. Turner, 1938:93, pl. 18, fig. 18. Vokes, 1939:150.

Galeodea (*Gomphopages*) *susanae* Schenck. Durham, 1942:184.

Hypotype: UCR 4683/15, locality 4683.

Local occurrence: Turritella uvasana applinae fauna, Juncal Formation; doubtfully identified in *Ectinochilus supraplicatus* fauna (UCR locality 4706).

Provincial Range: Early Eocene ("Capay Stage")? to late middle Eocene ("Transition Stage")?

A single specimen of this species, heavily encrusted with hard limestone matrix, was collected from locality 4683. A deformed specimen (UCR 4706/31) from locality 4706 also appears to represent this species but is too poorly preserved for positive identification.

Galeodea susanae is closely related to *G. sutterensis,* but is distinguished by larger size and by the presence of only two nodose carinae on the body whorl. The spinose nodes on the shoulder of the whorl of *G. susanae* are also longer than those of *G. sutterensis*. These two species appear to form an evolutionary sequence. They occur in stratigraphic sequence in the Pine Mountain section, in the Lodo Formation north of Coalinga, California (Vokes, 1939:33, 149), and in the Umpqua Formation in southern Oregon (Turner, 1938, tables 2 and 4).

Genus *Coalingodea* Durham, 1942

Type species: (by monotypy) *Galeodea tuberculiformis* Hanna.

Coalingodea tuberculiformis (Hanna)

(Pl. 8, fig. 7)

Morio (*Sconsia*) *tuberculatus* Gabb, 1864:104, pl. 19, fig. 57.

Galeodea tuberculata (Gabb). Dickerson, 1916:433, pl. 42, fig. 2.

Galeodea tuberculiformis Hanna, 1924:167 [new name for *Morio* (*Sconsia*) *tuberculatus* Gabb, a homonym of *Cassidaria tuberculata* Risso, 1826]. Schenck, 1926:83, 84, pl. 14, figs. 12–16. Stewart, 1926:380, 381, pl. 28, fig. 11. Vokes, 1939:149, 150, pl. 19, figs. 19, 21, 23–27.
Coalingodea tuberculiformis (Hanna). Durham, 1942:186, pl. 29, figs. 5, 9.
Cassis (*Coalingodea*) *tuberculata* (Gabb). Abbott, 1968:59, 60, pl. 34.

Hypotype: UCR 4752/1, locality 4752.

Local occurrence: *Turritella uvasana applinae* and *Ectinochilus supraplicatus* faunas, Juncal Formation.

Provincial Range: Middle Eocene ("Domengine" and "Transition" "Stages").

This distinctive species is characterized by its denticulate outer lip, beaded spiral sculpture, low spire, and one or more varices on the spire. The denticles on the outer lip and the beaded spiral sculpture readily distinguish it from all other cassid species described from the Lower Tertiary of the Pacific Coast.

Clark and Vokes (1936:861, fig. 1) and Vokes (1939:33) considered this species to be an index fossil for the "Domengine Stage." In the Pine Mountain section, however, it ranges up into the "Transition Stage." It has its lowest known occurrence in the "Domengine Stage."

Abbott (1968:59) considered *Coalingodea* Durham to be a subgenus of *Cassis* Scopoli. I prefer to regard it as a distinct genus.

Genus *Echinophoria* Sacco, 1890

Type species: (by subsequent designation, Dall, 1909) *Buccinum intermedium* Brocchi.

Echinophoria cf. *E. trituberculata* (Weaver)

Hypotype: UCR 4741/50, locality 4741.

Local occurrence: *Ectinochilus canalifer* fauna, Matilija Sandstone.

A few poorly preserved specimens comparable with Weaver's (1912:39,40, pl. 3, fig. 35; 1942:404, 405, pl. 78, figs. 10–15; pl. 79, figs. 1–4, 8) species were collected from localities 4715, 4738, and 4741 in the Matilija Sandstone. All are characterized by three rows of nodes on the body whorl, a high spire, and fine spiral lines that are not beaded. UCR 4741/50 shows a "corded" sutural collar which Durham (1942:185) considered to be characteristic of *Echinophoria*.

Echinophoria? sp.

Hypotype: UCR 4702/14, locality 4702.

Local occurrence: *Turritella uvasana applinae* fauna, Juncal Formation.

A poorly preserved specimen from locality 4702 is questionably referred to this genus. It has a high spire, three rows of nodes on the body whorl, and fine spiral lines that are not beaded. It lacks a "corded" sutural collar, however. The anterior canal is broken.

Family Cymatiidae

Genus *Sassia* Bellardi, 1872

Type species: (by original designation) *Triton apenninica* Sassi.

Sassia bilineata (Dickerson)

(Pl. 11, figs. 8, 10)

Fasciolaria bilineata Dickerson, 1916:493, pl. 37, figs. 6*a*, 6*b*. Hanna, 1927:319.
Sassia bilineata (Dickerson). Turner, 1938:91, pl. 18, fig. 20.

Hypotypes: UCR 4661/7, locality 4661; UCR 4683/12, locality 4683; UCR 4707/23, locality 4707.

Local occurrence: Turritella uvasana infera, T. uvasana applinae, and *Ectinochilus supraplicatus* faunas, Juncal Formation.

Provincial Range: Early Eocene ("Capay Stage") to late middle Eocene ("Transition Stage").

Family Bursidae

Genus *Olequahia* Stewart, 1926

Type species: (by original designation) *Cassidaria washingtoniana* Weaver.

Olequahia domenginica (Vokes)

(Pl. 9, figs. 4, 5)

Ranella domenginica Vokes, 1939:147, 148, pl. 19, figs. 6, 20.
Olequahia hornii domenginica (Vokes). Stewart, 1946, table 1.

Hypotypes: UCR 4679/1, locality 4679; UCR 4686/8, locality 4686.

Local occurrence: Turritella uvasana applinae fauna, Juncal Formation.

Provincial Range: Middle Eocene ("Domengine Stage").

Several poorly preserved specimens showing the characteristic ornamentation of this species were collected from localities 4679 and 4686.

Olequahia hornii (Gabb)

Tritonium hornii Gabb, 1864:94, pl. 28, fig. 208.
Bursa hornii (Gabb). Anderson and Hanna, 1925:54, 55, pl. 13, figs. 3, 4, 8.
Olequahia hornii (Gabb). Stewart, 1926:382, 383, pl. 29, figs. 1, 4, 18.

Hypotype: UCR 4726/20, locality 4726.

Local occurrence: Ectinochilus canalifer fauna, Matilija Sandstone.

Provincial Range: Late Eocene ("Tejon Stage"). This species has also been questionably identified from strata referable to the "Transition Stage" in western Santa Barbara County, California (Kleinpell and Weaver, 1963:191, pl. 26, fig. 6).

A single specimen was collected from locality 4726 in the Matilija Sandstone.

Genus *Ranellina* Conrad, 1865

Type species: (by original designation) *Ranellina maclurii* Conrad.

Ranellina pilsbryi Stewart

(Pl. 9, fig. 12)

Ranellina pilsbryi Stewart, 1926:384, 385, pl. 30, figs. 8, 9. Turner, 1938:91, pl. 16, fig. 3. Vokes, 1939:148, pl. 19, figs. 10, 17.

Hypotypes: UCR 4707/4, locality 4707; UCR 4721/112, locality 4721.

Local occurrence: Ectinochilus supraplicatus and *E. canalifer* faunas, Juncal and Matilija formations.

Provincial Range: Middle Eocene ("Domengine Stage") to late Eocene ("Tejon Stage").

This species is abundant in the *Ectinochilus canalifer* fauna. A few specimens were collected from the *E. supraplicatus* fauna.

Family Ficidae

Genus *Ficus* Röding, 1798

Type species: (by subsequent designation, Dall, 1906) *Ficus communis* Röding (= *Murex ficus* Linné).

Ficus mamillata Gabb

(Pl. 9, fig. 2)

Ficus mamillatus Gabb, 1864:211, pl. 32, fig. 276. Dickerson, 1915:88, pl. 6, fig. 12. Stewart, 1926:371, 372, pl. 29, fig. 12. Kleinpell and Weaver, 1963:191, pl. 26, fig. 3.

Ficus mamillata Gabb. Schenck and Keen, 1940, pl. 26, fig. 4. Keen and Bentson, 1944:159.
Hypotype: UCR 4715/9, locality 4715.
Local occurrence: Ectinochilus canalifer fauna, Matilija Sandstone.
Provincial Range: Late middle Eocene ("Transition Stage") ? to late Eocene ("Tejon Stage").

A single specimen was collected from locality 4715.

Genus *Ficopsis* Conrad, 1866

Type species: (by subsequent designation, Stewart, 1926) *Hemifusus remondii* Gabb.

Ficopsis cooperiana Stewart

(Pl. 9, figs. 7, 9)

Fusus (Hemifusus) cooperii Gabb, 1864:86, pl. 28, fig. 207.
Ficopsis cooperii (Gabb). Dickerson, 1915:61, 62, pl. 6, fig. 11; 1916:492, pl. 37, fig. 7.
Ficopsis cooperiana Stewart, 1926:378, 379 [new name for *Fusus (Hemifusus) cooperii* Gabb, preoccupied].
Hypotypes: UCR 4679/111, locality 4679; UCR 4703/14, locality 4703.
Local occurrence: Turritella uvasana applinae fauna, Juncal Formation.
Provincial Range: Middle Eocene ("Domengine Stage").

This species was collected from a number of localities in the *Turritella uvasana applinae* fauna. It is characterized by three rows of irregularly spaced spinose nodes on the body whorl and by a biangulate spire. The middle row of nodes is closer to the posterior than to the anterior row. The presutural shelf is concave in profile.

Ficopsis hornii (Gabb)

(Pl. 9, fig. 10)

Fusus (Hemifusus) hornii Gabb, 1864:86, pl. 28, figs. 206, 206*a*.
Ficopsis hornii (Gabb). Dickerson, 1915: 61, pl. 6, fig. 9. Anderson and Hanna, 1925:111, pl. 12, fig. 8. Stewart, 1926:377, 378, pl. 30, figs. 3, 4. Kleinpell and Weaver, 1963:191, pl. 26, fig. 1.
Hypotypes: UCR 4706/8, locality 4706; UCR 4738/18, locality 4738; UCR 4741/201, locality 4741.
Local occurrence: Ectinochilus supraplicatus and *E. canalifer* faunas, Juncal and Matilija formations.
Provincial Range: Late middle Eocene ("Transition Stage") to late Eocene (*Turritella schencki delaguerrae* Zone of Kleinpell and Weaver, 1963).

This species resembles *Ficopsis cooperiana* in having three rows of nodes on the body whorl, but is distinguished by the alignment of the nodes along prominent axial ribs. The spire is unicarinate, the presutural shelf is flat to very gently concave, and the middle row of nodes is situated closer to the anterior rather than to the posterior row.

Ficopsis remondii (Gabb) s.s.

Fusus (Hemifusus) remondii Gabb, 1864:87, pl. 18, fig. 36.
Ficopsis remondii (Gabb). Dickerson, 1915:61, pl. 6, fig. 8. Stewart, 1926:376, pl. 30, figs. 1, 2. Turner, 1938:92, 93, pl. 15, fig. 20. Kleinpell and Weaver, 1963:191, pl. 26, fig. 2.
Hypotypes: UCR 4714/13, locality 4714; UCR 4741/1, locality 4741.
Local occurrence: Ectinochilus canalifer fauna, Matilija Sandstone.
Provincial Range: Late Eocene ("Tejon Stage" and *Turritella schencki delaguerrae* Zone of Kleinpell and Weaver, 1963).

Poorly preserved specimens of this species were collected from several localities in the Matilija Sandstone. Typical *Ficopsis remondii* is characterized by its rounded to faintly tricarinate body whorl.

Ficopsis remondii crescentensis Weaver and Palmer

(Pl. 9, fig. 11)

Ficopsis remondii (Gabb) var. *crescentensis* Weaver and Palmer, 1922:p. 39, 40, pl. 11, fig. 14. Turner, 1938:93, pl. 15, fig. 19. Weaver, 1942:399, pl. 77, fig. 10.

Ficopsis remondii (Gabb) *crescentensis* Weaver and Palmer. Vokes, 1939:p. 152, 153.

Hypotypes: UCR 4667/8, locality 4667; UCR 4703/13, locality 4703; UCR 4707/30, locality 4707.

Local occurrence: Turritella uvasana infera, T. uvasana applinae, and *Ectinochilus supraplicatus* faunas, Juncal Formation.

Provincial Range: Early Eocene ("Capay Stage") to late middle Eocene ("Transition Stage").

This subspecies is distinguished from the typical subspecies by the presence of three prominent carinae on the body whorl. It has its highest-known occurrence in the "Transition Stage" and appears to have evolved into the typical subspecies by progressive loss of the carinae.

Superfamily *MURICACEA*

Family Muricidae

Subfamily Muricinae

Genus *Pterynotus* Swainson, 1833

Type species: (by original designation) *Murex pinnatus* Wood.

Pterynotus n. sp.?

Hypotypes: UCR 4676/31, locality 4676; UCR 4683/16, locality 4683; UCR 4702/13, locality 4702.

Local occurrence: Turritella uvasana applinae fauna, Juncal Formation.

Three poorly preserved specimens of *Pterynotus* collected from the *Turritella uvasana applinae* fauna probably represent a new species. This genus has not been previously reported from the Eocene of the Pacific Coast. Each specimen contains three winged varices that are spirally continuous up the spire and along the anterior canal and alternate with three axial rows of low, blunt nodes. Faint widely spaced spiral lines are visible on the whorls of two of the specimens.

Genus *Laevityphis* Cossmann, 1903

Type species: (by original designation) *Typhis coronarius* Deshayes.

Subgenus *Laevityphis* s.s.

Laevityphis (*Laevityphis*) *antiquus* (Gabb)

Typhis antiquus Gabb, 1864:82, pl. 18, fig. 31; 1869:214. Stewart, 1926:387, 388, pl. 27, figs. 7, 8. Weaver, 1953:43.

Laevityphis (*Laevityphis*) *antiquus* (Gabb). Keen, 1944:58, 63.

Hypotypes: UCR 4690/10, locality 4690; UCR 4707/24, locality 4707; UCR 4731/15, locality 4731.

Local occurrence: Turritella uvasana applinae, Ectinochilus supraplicatus, and *E. canalifer* faunas, Juncal and Matilija formations.

Provincial Range: Middle Eocene ("Domengine Stage") to late Eocene ("Tejon Stage").

Three specimens of this species were collected in the Pine Mountain section.

Genus *Hexaplex* Perry, 1810

Type species: (by subsequent designation, Iredale, 1915) *Hexaplex foliacea* Perry (= *Murex chichoreum* Gmelin).

Hexaplex? whitneyi (Gabb)

(Pl. 9, fig. 13)

Tritonium whitneyi Gabb, 1864:96, pl. 28, figs. 210, 210*a*.
Murex beali Anderson and Hanna, 1925:50, 51, pl. 13, fig. 16.
Murex whitneyi (Gabb). Stewart, 1926:387, pl. 30, fig. 10. Kleinpell and Weaver, 1963:192, 193, pl. 26, fig. 11.

Hypotype: UCR 4715/3, locality 4715.
Local occurrence: Ectinochilus canalifer fauna, Matilija Sandstone.
Provincial Range: Late middle Eocene ("Transition Stage") to late Eocene ("Tejon Stage").

This species is represented in the *Ectinochilus canalifer* fauna by a single specimen. Its generic position is uncertain. It lacks the rounded body whorl and long spinose anterior canal characteristic of *Murex* s.s. I am following E. H. Vokes (1971:117) in questionably referring it to *Hexaplex*.

The specimens identified by H. E. Vokes (1939:144, pl. 19, figs. 7, 12) as *Muricopsis* (?) *whitneyi* (Gabb) var. from the type Domengine Formation are not closely related to *Hexaplex? whitneyi*. They have a much shorter anterior canal, a narrower presutural shelf, and the principal spiral ribs are stronger and more widely spaced. The Domengine specimens have denticles on the outer lip but lack the small denticles near the base of the columella characteristic of *Muricopsis*. They may be referable to *Hexaplex*. They resemble *Hexaplex katherinae* E. H. Vokes (1968:100, pl. 1, figs. 4*a*, 4*b*)

Superfamily BUCCINACEA

Family Buccinidae

Genus *Siphonalia* A. Adams, 1863

Type species: (by subsequent designation, Cossmann, 1889) *Buccinum cassidariaeformis* Reeve.

Siphonalia sopenahensis (Weaver)

(Pl. 9, fig. 8)

Hemifusus sopenahensis Weaver, 1912:44, 45, pl. 1, figs. 2, 3. Dickerson, 1915:37, pl. 8, figs. 2*a*, 2*b*.
Tritonium sopenahensis Weaver, 1912:40, pl. 1, fig. 6.
Siphonalia sopenahensis (Weaver). Turner, 1938:31, 35. Weaver, 1942:437, 438, pl. 86, figs. 1–8, 15.
Siphonalia (*Nassicola*) *sopenahensis* (Weaver). Ruth, 1941:291, 292, pl. 47, figs. 5, 6.

Hypotype: UCR 4719/16, locality 4719.
Local occurrence: Ectinochilus canalifer fauna, Matilija Sandstone.
Provincial Range: Late Eocene ("Tejon Stage").

Siphonalia cf. *S. thunani* (Dickerson)

Hypotype: UCR 4661/14, locality 4661.
Local occurrence: Turritella uvasana infera fauna, Juncal Formation.

A single poorly preserved specimen comparable with Dickerson's (1916:496, pl. 41, figs. 8*a*, 8*b*) species was collected from locality 4661.

Genus *Brachysphingus* Gabb, 1869

Type species: (by subsequent designation, Cossmann, 1901) *Brachysphingus sinuatus* Gabb.

Brachysphingus mammilatus Clark and Woodford

(Pl. 10, fig. 3)

Brachysphingus mammilatus Clark and Woodford, 1927:116, 117, pl. 20, figs. 8–15.
Hypotype: UCR 4662/5, locality 4662.
Local occurrence: Turritella uvasana infera fauna, Juncal Formation.
Provincial Range: Late Paleocene ("Meganos Stage") to early Eocene ("Capay Stage").

A single slightly deformed specimen of this species was collected from locality 4662.

Genus *Molopophorus* Gabb, 1869

Type species: (by monotypy) *Bulla* (*Molopophorus*) *striata* Gabb.

Molopophorus antiquatus (Gabb)

(Pl. 10, fig. 1)

Nassa antiquata Gabb, 1864:97, pl. 18, fig. 50.
Molopophorus antiquatus (Gabb). Stewart, 1926:390, 391, pl. 28, fig. 4. Turner, 1938:77, pl. 15, fig. 11. Vokes, 1939:142, pl. 19, figs. 1–3.
Hypotype: UCR 4707/301, locality 4707.
Local occurrence: Ectinochilus supraplicatus fauna, Juncal Formation.
Provincial Range: Early Eocene ("Capay Stage") to late middle Eocene ("Transition Stage").

This species is fairly common in the *Ectinochilus supraplicatus* fauna. It is characterized by the possession of numerous fine closely spaced axial and spiral ribs. The axial ribs are more prominent than the spiral ribs.

Molopophorus cretaceus (Gabb) closely resembles *M. antiquatus,* but is distinguished by fewer and more widely spaced axial ribs on the adult whorls and by a subsutural collar with two prominent spiral ribs rather than the four weak spiral ribs on the collar of *M. antiquatus.*

Molopophorus antiquatus has its highest known occurrence in the "Transition Stage." It may be ancestral to *M. tejonensis* of the "Tejon Stage."

Molopophorus tejonensis Dickerson

(Pl. 10, fig. 2)

Molopophorus tejonensis Dickerson, 1915:66, 67, pl. 8, figs. 3*a*, 3*b*. Clark, 1938:715, pl. 4, figs. 38, 39, 47.
Cominella tejonensis (Dickerson). Anderson and Hanna, 1925:72.
Hypotype: UCR 4722/401, locality 4722.
Local occurrence: Ectinochilus canalifer fauna, Juncal and Matilija formations.
Provincial Range: Late Eocene ("Tejon Stage").

Abundant well-preserved specimens of this species were collected from several localities in the *Ectinochilus canalifer* fauna. It is characterized by the possession of axial and spiral ribs of approximately equal strength and spacing, forming a markedly reticulate sculpture. Small nodes are formed at the intersection of the axial and spiral ribs.

Stewart (1926:392) considered *Molopophorus tejonensis* to be a synonym of *M. cretaceus* (Gabb). As pointed out by Clark (1938: 716), however, the two species are quite distinct. *M. cretaceus* contains fewer and more widely spaced axial ribs than *M. tejonensis* and the axial ribs are more prominent than the spiral ribs. The subsutural collar is also better defined on *M. cretaceus. Molopophorus tejonensis* is distinguished from *M. antiquatus* by more prominent and widely spaced axial and spiral ribbing.

Family Fasciolariidae
Subfamily Fasciolariinae
Genus *Clavilithes* Swainson, 1840

Type species: (by subsequent designation, Grabau, 1904) ***Fusus longaevus*** Deshayes, non Solander (= ***Fusus parisiensis*** Mayer-Eymar).

Clavilithes tabulatus (Dickerson)

(Pl. 10, figs. 4, 5)

Clavella tabulata Dickerson, 1913:283: pl. 12, fig. 7.
Clavilithes tabulatus (Dickerson). Clark and Vokes, 1936:874, pl. 1, fig. 3 [holotype refigured].
?Clavilithes cf. ***C. tabulatus*** (Dickerson). Crowell and Susuki, 1959:588, pl. 2, figs. 6, 7.

Hypotypes: UCR 4658/7, locality 4658; UCR 4661/5, locality 4661.
Local occurrence: Turritella uvasana infera fauna, Juncal Formation.
Provincial Range: Early Eocene ("Capay Stage").

Two incomplete specimens of this species were collected from the *Turritella uvasana infera* fauna. Except for the presence of a narrow channel along the suture, they agree closely with the holotype.

The specimen figured by Crowell and Susuki (1959) from the lower part of the Maniobra Formation in eastern Riverside County, California, appears to be referable to this species.

Subfamily Fusininae
Genus *Fusinus* Rafinesque, 1815

Type species: (by monotypy) ***Murex colus*** Linné.

Fusinus teglandae Hanna

(Pl. 10, figs. 6, 7)

Fusinus teglandi Hanna, 1927:315, pl. 51, fig. 9.
Fusinus teglandae Hanna. Keen and Bentson, 1944: 161.

Hypotypes: UCR 4684/5, locality 4684; UCR 4700/4, locality 4700.
Local occurrence: Turritella uvasana applinae fauna, Juncal Formation.
Provincial Range: Middle Eocene ("Domengine Stage").
Two broken specimens of this species were collected from the ***T. uvasana applinae*** fauna.

Genus *Falsifusus* Grabau, 1904

Type species: (by original designation) ***Fusus ottonis*** Aldrich.

Falsifusus cf. *F. marysvillensis* (Merriam and Turner)

(Pl. 10, fig. 8)

Hypotype: UCR 4659/9, locality 4659.
Local occurrence: Turritella uvasana infera fauna, Juncal Formation.

A single broken specimen comparable with this species was collected from locality 4659.

Genus *Perse* Clark, 1918 (= *Whitneyella* Stewart, 1926)

Type species: (by original designation) ***Perse corrugatum*** Clark.

Perse martinez (Gabb)

(Pl. 10, fig. 9)

Fusus martinez Gabb, 1864:82, pl. 18, fig. 32.

Whitneyella martinez (Gabb). Stewart, 1926:403, 404, pl. 26, fig. 7.
Falsifusus martinez (Gabb). Weaver, 1953:43.
Hypotype: UCR 4707/201, locality 4707.
Local occurrence: Ectinochilus supraplicatus fauna, Juncal Formation.
Provincial Range: Middle Eocene ("Domengine" and "Transition" "Stages").

This species is abundant in the *E. supraplicatus* fauna.

Perse sinuata (Gabb) s.s.

(Pl. 10, fig. 13)

Fasciolaria sinuata Gabb, 1864:101, pl. 28, figs. 213, 213*a*. Dickerson, 1915:74, pl. 11, figs. 3*a*, 3*b*.
Latirus sinuatus (Gabb). Anderson and Hanna, 1925:64, 65, pl. 13, figs. 14, 15.
Whitneyella sinuata (Gabb). Stewart, 1926:401, 402, pl. 28, figs. 6, 17.
Hypotype: UCR 4719/2, locality 4719.
Local occurrence: Ectinochilus canalifer fauna, Matilija Sandstone.
Provincial Range: Late Eocene ("Tejon Stage").

A single well-preserved specimen was collected from locality 4719.

Superfamily VOLUTACEA

Family OLIVIDAE

Subfamily PSEUDOLIVINAE

Genus *Pseudoliva* Swainson, 1840

Type species: (by original designation) *Buccinum plumbea* Chemnitz (=*Buccinum crassa* Gmelin).

Pseudoliva inornata Dickerson

(Pl. 9, fig. 6)

Pseudoliva inornata Dickerson, 1915:62, 63, pl. 7, fig. 1*a* [not figs. 1*b*, 1*c*, =*Pseudoliva kirbyi* Clark, 1938:709]. Anderson and Hanna, 1925:52, pl. 12, fig. 1.
Pseudoliva tejonensis Dickerson, 1915:63, pl. 7, fig. 2.
Hypotypes: UCR 4707/31, locality 4707; UCR 4738/6, locality 4738; UCR 4741/14, locality 4741.
Local occurrence: Ectinochilus supraplicatus and *E. canalifer* faunas, Juncal and Matilija formations.
Provincial Range: Late middle Eocene ("Transition Stage") to late Eocene ("Tejon Stage").

A few specimens of this species were collected from the *E. supraplicatus* and *E. canalifer* faunas. Specimen 4741/14 contains numerous fine spiral lines on the body whorl and represents the form described by Dickerson (1915) as *Pseudoliva tejonensis,* which Anderson and Hanna (1925:52) regarded as a synonym of *P. inornata.*

Pseudoliva volutaeformis Gabb

(Pl. 9, fig. 14)

Pseudoliva volutaeformis Gabb, 1864:99, pl. 28, fig. 212. Dickerson, 1915:90, pl. 7, figs. 3*a*, 3*b*. Stewart, 1926:400, pl. 29, fig. 7.
Hypotype: UCR 4706/151, locality 4706.
Local occurrence: Ectinochilus supraplicatus fauna, Juncal Formation.
Provincial Range: Late middle Eocene ("Transition Stage") to late Eocene ("Tejon Stage").

Two well-preserved specimens of this species were collected from locality 4706.

Genus *Strepsidura* Swainson, 1840

Type species: (by original designation) *Strepsidura costata* Swainson.

Strepsidura ficus (Gabb)

(Pl. 10, fig. 10)

Whitneya ficus Gabb, 1864:104, pl. 28, fig. 216. Dickerson, 1915, p. 69, pl. 9, figs. *5a–5d*.
Strepsidura ficus (Gabb). Stewart, 1926:404, 405, pl. 29, fig. 11. Kleinpell and Weaver, 1963:193, pl. 27, figs. 1–3.

Hypotypes: UCR 4683/6, locality 4683; UCR 4708/50, locality 4708; UCR 4719/51, locality 4719; UCR 4721/131, locality 4721.

Local occurrence: Turritella uvasana applinae, Ectinochilus supraplicatus, and *E. canalifer* faunas, Juncal and Matilija formations.

Provincial Range: Middle Eocene ("Domengine Stage") to late Eocene ("Tejon Stage").

This species is abundant in the *Ectinochilus canalifer* fauna. A few specimens were collected from the *E. supraplicatus* fauna and one from the *Turritella uvasana applinae* fauna.

Subfamily Olivinae

Genus *Olivella* Swainson, 1831

Type species: (by subsequent designation, Dall, 1909) *Oliva purpurata* Swainson (= *Oliva dama* Mawe).

Olivella mathewsonii Gabb

Olivella mathewsonii Gabb, 1864:100, pl. 18, fig. 53. Stewart, 1926:410, pl. 29, fig. 13.

Hypotypes: UCR 4688/23, locality 4688; UCR 4708/18, locality 4708; UCR 4717/7, locality 4717.

Local occurrence: Turritella uvasana applinae, Ectinochilus supraplicatus, and *E. canalifer* faunas, Juncal and Matilija formations.

Provincial Range: Middle Eocene ("Domengine Stage") to late Eocene ("Tejon Stage").

Family Volutomitridae

Genus *Proximitra* Finlay, 1927

Type species: (by original designation) *Vexillum rutidolomum* Suter.

Proximitra? cretacea (Gabb)

Mitra cretacea Gabb, 1864:103, pl. 28, fig. 215. Stewart, 1926:406, pl. 27, figs. 9, 10.
Uromitra (?) *cretacea* (Gabb). Vokes, 1939:134, 135, pl. 18, fig. 19.
Dentimitra cretacea (Gabb). Cernohorsky, 1970:41.

Hypotypes: UCR 4676/30, locality 4676; UCR 4686/7, locality 4686.

Local occurrence: Turritella uvasana applinae fauna, Juncal Formation.

Provincial Range: Middle Eocene ("Domengine Stage").

This species is characterized by the presence of a prominent nodose carina on the shoulder of the whorls. The surface of the whorls is also covered by numerous fine spiral lines. There are four folds on the columella, the first posterior fold is shorter and slightly less prominent than the second posterior fold.

The generic position of this species is uncertain. Cernohorsky (1970:41) recently assigned it to *Dentimitra* Koenen, 1890. In that genus, however, the first posterior fold on the columella is longer and more prominent than the second fold (see Cernohorsky, 1970:40, 41). The presence of the prominent nodose carina on the shoulder of the whorls and the stronger second posterior fold on the columella suggest that Gabb's species is more closely related to *Proximitra* and it is therefore questionably referred to that genus.

Genus *Conomitra* Conrad, 1865

Type species: (by subsequent designation, Fischer, 1884) *Mitra fusoides* Lea.

Conomitra aff. *C. washingtoniana* (Weaver)

(Pl. 10, fig. 14)

Hypotype: UCR 4706/26, locality 4706.
Local occurrence: Ectinochilus supraplicatus fauna, Juncal Formation.

A single specimen collected from locality 4706 closely resembles Weaver's (1912:52, pl. 2, fig. 16; 1942:497, 498, pl. 95, figs. 8, 9, 16) species, but is distinguished by stronger spiral sculpture and more prominent axial ribs at the posterior margin of the whorls below the suture. These differences may be of taxonomic significance because the Pine Mountain specimen occurs in older strata ("Transition Stage") than Weaver's species ("Tejon Stage").

Family Vasidae

Genus *Pseudoperissolax* Clark, 1918

Type species: (by original designation) *Busycon? blakei* Conrad.

Pseudoperissolax blakei praeblakei Vokes

(Pl. 10, figs. 15, 16)

Perissolax blakei (Conrad). Gabb, 1864:92 [in part], pl. 21, fig. 110.
Pseudoperissolax blakei (Conrad) (subsp.?). Stewart, 1926:429, pl. 28, fig. 1.
Pseudoperissolax blakei (Conrad) subsp. *praeblakei* Vokes, 1939:145, pl. 19, figs. 14, 22.
Hypotypes: UCR 4682/7, locality 4682; UCR 4685/121, locality 4685.
Local occurrence: Turritella uvasana applinae fauna, Juncal Formation.
Provincial Range: Early Eocene ("Capay Stage") to middle Eocene ("Domengine Stage").

Poorly preserved specimens of this subspecies were collected from several localities in the *Turritella uvasana applinae* fauna. *P. blakei praeblakei* is distinguished from the typical subspecies by a higher spire, a flatter shoulder on the whorl, and finer spiral ornamentation (compare figs. 15 and 17 on pl. 11).

Pseudoperissolax aff. *P. blakei* (Conrad) s.s.

Hypotype: UCR 4708/7, locality 4708.
Local occurrence: Ectinochilus supraplicatus fauna, Juncal Formation.

One young adult and several immature specimens of a *Pseudoperissolax* were collected from the *E. supraplicatus* fauna. The adult specimen resembles *P. blakei* s.s. in having relatively coarse spiral lines on the whorl, but is distinguished from it by a nearly flat shoulder like that of *P. blakei praeblakei*. The specimen appears to be intermediate between the two subspecies.

Family Volutidae

Subfamily Volutinae

Genus *Volutocristata* Gardner and Bowles, 1934

Type species: (by original designation) *Volutocristata chiapasensis* Gardner and Bowles.

Volutocristata lajollaensis (Hanna)

Pejona[3] *lajollaensis* Hanna, 1927:320, pl. 42, figs. 1, 2.
Volutocristata lajollaensis (Hanna). Gardner and Bowles, 1934:246, fig. 13.
Hypotypes: UCR 4699/13, locality 4699.

[3] Error for *Plejona*.

Local occurrence: *Turritella uvasana applinae* fauna, Juncal Formation.
Provincial Range: Middle Eocene ("Domengine Stage").

A single broken specimen of this distinctive species was collected from locality 4699.

Genus *Cryptochorda* Mörch, 1858

Type species: (by monotypy) *Buccinum stromboides* Hermann.

Subgenus *Cryptochorda* s.s.

Cryptochorda (*Cryptochorda*) cf. *C.* (*C.*) *californica* (Cooper)

Hypotype: UCR 4668/10, locality 4668.
Local occurrence: *Turritella uvasana infera* fauna, Juncal Formation.

Two poorly preserved broken specimens comparable with this species were collected from locality 4668.

Family CANCELLARIIDAE

Genus *Bonellitia* Jousseaume, 1887

Type species: (by original designation) *Cancellaria bonellii* Bellardi.

Subgenus *Admetula* Cossmann, 1889

Type species: (by original designation) *Cancellaria evulsa* Solander.

Bonellitia (*Admetula*) *paucivaricata* (Gabb)

Tritonium paucivaricata Gabb, 1864:95, pl. 28, figs. 209, 209*a*.
Cancellaria paucivaricata (Gabb). Anderson and Hanna, 1925:81, pl. 8, figs. 3, 4.
Bonellitia (*Admetula*) *paucivaricata* (Gabb). Stewart, 1926:413, pl. 29, fig. 5. Turner, 1938:71, pl. 15, figs. 12, 13.
Hypotypes: UCR 4676/28, locality 4676; UCR 4703/40, locality 4703.
Local occurrence: *Turritella uvasana applinae* fauna, Juncal Formation.
Provincial Range: Early Eocene ("Capay Stage") to late Eocene ("Tejon Stage").

Superfamily CONACEA

Family TURRIDAE

Subfamily TURRICULINAE

Genus *Turricula* Schumacher, 1817

Type species: (by monotypy) *Turricula flammea* Schumacher.

Turricula cohni (Dickerson)

(Pl. 11, fig. 6)

Surcula cohni Dickerson, 1915:70, pl. 10, fig. 1. Anderson and Hanna, 1925:84, pl. 10, figs. 12–14; pl. 11, figs. 3, 4.
Turricula cohni (Dickerson). Keen and Bentson, 1944:196.
Hypotypes: UCR 4705/13, locality 4705; UCR 4741/251, locality 4741.
Local occurrence: *Ectinochilus supraplicatus* and *E. canalifer* faunas, Juncal and Matilija formations.
Provincial Range: Late middle Eocene ("Transition Stage") to late Eocene ("Tejon Stage").

A few well-preserved specimens of this species were collected from the *Ectinochilus supraplicatus* and *E. canalifer* faunas.

Turricula praeattenuata (Gabb)

(Pl. 11, figs. 4, 11)

Surcula praeattenuata Gabb, 1869:150, pl. 26, fig. 27. Hanna, 1927:326, pl. 54, figs. 2, 7, 9, 10.

Turricula praeattenuata (Gabb). Keen and Bentson, 1944:197.
Hypotypes: UCR 4658/13, locality 4658; UCR 4676/4, locality 4676; UCR 4686/6, locality 4686.
Local occurrence: Turritella uvasana infera and *T. uvasana applinae* faunas, Juncal Formation.
Provincial Range: Early Eocene ("Capay Stage") to middle Eocene ("Domengine Stage").

Three specimens of this species were collected.

Genus *Pleurofusia* de Gregorio, 1890

Type species: (by original designation) *Pleurotoma* (*Pleurofusia*) *longirostropis* de Gregorio.

Pleurofusia fresnoensis (Arnold)

(Pl. 11, fig. 9)

Pleurotoma fresnoensis Arnold, 1910:53, pl. 4, fig. 23.
Pleurofusia fresnoensis (Arnold). Vokes, 1939:117, pl. 17, figs. 15, 16.
Hypotypes: UCR 4750/30, locality 4750; UCR 4706/231, locality 4706.
Local occurrence: Turritella uvasana applinae and *Ectinochilus supraplicatus* faunas, Juncal Formation.
Provincial Range: Middle Eocene ("Domengine" and "Transition" "Stages").

This species was considered by Vokes (1939:35) to be an index fossil for the "Domengine Stage." In the Pine Mountain section, however, its range extends into the "Transition Stage."

Genus *Eosurcula* Casey, 1904

Type species: (by subsequent designation, Vokes, 1939) *Surcula moorei* Gabb.

Eosurcula capayana Vokes

Eosurcula capayana Vokes, 1939:188, pl. 17, figs. 12, 13.
Hypotype: UCR 4661/6, locality 4661.
Local occurrence: Turritella uvasana infera fauna, Juncal Formation.
Provincial Range: Early Eocene ("Capay Stage").

A single broken specimen showing the three prominent equally spaced primary carinae on the body whorl characteristic of this species was collected from locality 4661. Only the posterior two primaries are exposed on the spire whorls.

Genus *Surculites* Conrad

Type species: (by monotypy) *Surcula* (*Surculites*) *annosus* Conrad.

Surculites mathewsonii (Gabb)

(Pl. 11, figs. 5, 7)

Fusus mathewsonii Gabb, 1864:83, pl. 18, fig. 33.
Potamides? davisiana Cooper, 1894:44, pl. 1, fig. 13.
Surcula davisiana (Cooper). Dickerson, 1913:279, pl. 12, figs. 6*a*, 6*b*.
Pleurotoma decipiens Cooper, 1894:40, pl. 2, fig. 32.
Surculites mathewsonii (Gabb). Stewart, 1926:420, pl. 26, figs. 12–14. Turner, 1938:69, pl. 17, figs. 6, 10. Vokes, 1939:123, pl. 17, figs. 8, 19.
Surcula decipiens (Cooper). Hanna, 1927:324, pl. 54, figs. 6, 8.
Hypotypes: UCR 4661/15, locality 4661; UCR 4679/4, locality 4679; UCR 4682/2, locality 4682.
Local occurrence: Turritella uvasana infera and *T. uvasana applinae* faunas, Juncal Formation.
Provincial Range: Early Eocene ("Capay Stage") to middle Eocene ("Domengine Stage").

Genus *Nekewis* Stewart, 1926

Type species: (by original designation) *Fasciolaria washingtoniana* Weaver.

Nekewis io (Gabb)

(Pl. 11, fig. 3)

?*Fasciolaria io* Gabb, 1864:101, pl. 28, fig. 214.
Surcula io (Gabb). Dickerson, 1915:72, pl. 10, fig. 11. Anderson and Hanna, 1925:88, pl. 10, fig. 11.
Surcula ioformis Anderson and Hanna, 1925:89, pl. 12, figs. 3, 7.
Surcula alizensis Anderson and Hanna, 1925:89, pl. 12, figs. 2, 4.
Nekewis io (Gabb). Stewart, 1926:421, pl. 30, fig. 11. Clark, 1938:723, pl. 2, figs. 25, 26.

Hypotypes: UCR 4706/4, locality 4706; UCR 4708/1, locality 4708; UCR 4741/11, locality 4741.

Local occurrence: Ectinochilus supraplicatus and *E. canalifer* faunas, Juncal and Matilija formations.

Provincial Range: Late middle Eocene ("Transition Stage") to late Eocene ("Tejon Stage").

Three specimens of this distinctive species were collected, two in the *Ectinochilus supraplicatus* fauna and one in the *E. canalifer* fauna.

Subfamily Turrinae

Genus *Gemmula* Weinkauff, 1876

Type species: (by subsequent designation, Cossmann, 1896) *Pleurotoma gemmata* Hinds.

Gemmula abacta Anderson and Hanna

Gemmula abacta Anderson and Hanna, 1925:92, pl. 7, figs. 6, 7; pl. 8, fig. 11.

Hypotypes: UCR 4721/29, locality 4721; UCR 4726/16, locality 4726.

Local occurrence: Ectinochilus canalifer fauna, Matilija Sandstone.

Provincial Range: Late Eocene ("Tejon Stage").

Genus *Hemipleurotoma* Cossmann, 1889

Type species: (by original designation) *Pleurotoma archimedis* Bellardi.

Hemipleurotoma pulchra (Dickerson)

Turris pulchra Dickerson, 1915:71, pl. 10, figs. 4*a*, 4*b*.
Hemipleurotoma pulchra (Dickerson). Turner, 1938:71, pl. 17, figs. 15, 17.

Hypotype: UCR 4728/6, locality 4728.

Local occurrence: Ectinochilus canalifer fauna, Matilija Sandstone.

Provincial Range: Late Eocene ("Tejon Stage").

Subfamily Brachytominae

Genus *Exilia* Conrad, 1860

Type species: (*by monotypy*) *Exilia pergracilis* Conrad.

Exilia microptygma (Gabb)

Cordiera microptygma Gabb, 1864:93, pl. 28, fig. 203.
Exilia novatrix Anderson and Hanna, 1925:60, pl. 7, fig. 10.
Exilia microptygma (Gabb). Stewart, 1926:418, pl. 29, fig. 10*a*. Bentson, 1940:212, pl. 1, figs. 25, 26; pl. 2, figs. 11, 13.

Hypotype: UCR 4706/22, locality 4706.

Local occurrence: Ectinochilus supraplicatus fauna, Juncal Formation.

Provincial Range: Late middle Eocene ("Transition Stage") to late Eocene ("Tejon Stage").

Exilia fausta Anderson and Hanna

(Pl. 11, fig. 2)

Exilia fausta Anderson and Hanna, 1925:59, pl. 8, figs. 8, 9. Bentson, 1940:214, pl. 3, fig. 14.

Hypotype: UCR 4726/19, locality 4726.
Local occurrence: *Ectinochilus canalifer* fauna, Matilija Sandstone.
Provincial Range: Late Eocene ("Tejon Stage").

Family Conidae

Genus *Conus* Linné, 1758

Type species: (by subsequent designation, Children, 1823) *Conus marmoreus* Linné.

Conus n. sp.? aff. *C. californianus* (Conrad)

(Pl. 11, fig. 1)

Hypotype: UCR 4706/141, locality 4706.
Local occurrence: *Ectinochilus supraplicatus* fauna, Juncal Formation.

Well-preserved specimens of *Conus* collected from locality 4706 closely resemble *C. californianus* (Conrad, 1855:11; 1857:322, pl. 2, fig. 9), but are distinguished by more numerous and closely spaced nodes. There are 14–16 nodes on the body whorl of the Pine Mountain specimens as compared with 10–12 on the body whorl of *C. californianus* from the type Tejon Formation. Whether this difference in number of nodes is of taxonomic significance is uncertain because the range of variation of *C. californianus* with respect to this character has not been determined. It is perhaps significant, however, that the Pine Mountain specimens occur in strata older ("Transition Stage") than typical *C. californianus* ("Tejon Stage").

Conus hornii Gabb s.s.

(Pl. 10, fig. 12)

Conus hornii Gabb, 1864:122, pl. 29, fig. 226. Dickerson, 1915:98, pl. 11, figs. 9*a*–9*c*. Anderson and Hanna, 1925:99. Stewart, 1926:415, pl. 29, fig. 16.
Hypotype: UCR 4731/14, locality 4731.
Local occurrence: *Ectinochilus canalifer* fauna, Matilija Sandstone.
Provincial Range: Late Eocene ("Tejon Stage").

Family Terebridae

Genus *Terebra* Bruguiere, 1789

Type species: (by subsequent designation, Lamarck, 1799) *Buccinum subulata* Linné.

Terebra californica Gabb

Terebra californica Gabb, 1869:162, pl. 27, fig. 41. Anderson and Hanna, 1925:82, pl. 8, fig. 18. Stewart, 1926:424, pl. 26, fig. 5. Vokes, 1939:113, pl. 16, fig. 3*a*.
Hypotypes: UCR 4705/10, locality 4705; UCR 4722/10, locality 4722.
Local occurrence: *Ectinochilus supraplicatus* and *E. canalifer* faunas, Juncal and Matilija formations.
Provincial Range: Middle Eocene ("Domengine Stage") to late Eocene ("Tejon Stage").

Superfamily ACTEONACEA

Family Acteonidae

Genus *Acteon* Montfort, 1810

Type species: (by monotypy) *Voluta tornatilis* Gmelin.

Acteon quercus Anderson and Hanna

(Pl. 10, fig. 11)

Acteon quercus Anderson and Hanna, 1925:141, pl. 8, fig. 1.

Hypotype: UCR 4722/14, locality 4722.
Local occurrence: Ectinochilus canalifer fauna, Juncal and Matilija formations.
Provincial Range: Late Eocene ("Tejon Stage").

Family Akeridae

Genus Akera Müller, 1776

Type species: (by original designation) *Akera bullata* Müller.

Akera maga Vokes

Akera maga Vokes, 1939:111, pl. 16, figs. 34, 40, 41.
Hypotypes: UCR 4656/4, locality 4656; UCR 4679/29, locality 4679.
Local occurrence: Turritella uvasana infera and *T. uvasana applinae* faunas, Juncal Formation.
Provincial Range: Early Eocene ("Capay Stage") to middle Eocene ("Domengine Stage").

Family Scaphandridae

Genus *Abderospira* Dall, 1895

Type species: (by original designation) *Bullina chipolana* Dall.

Abderospira hornii (Gabb)

Bulla hornii Gabb, 1864:143, pl. 29, fig. 325.
Abderospira hornii (Gabb). Stewart, 1926:439, pl. 29, fig. 9.
Hypotype: UCR 4722/22, locality 4722.
Local occurrence: Ectinochilus canalifer fauna, Juncal Formation.
Provincial Range: Late Eocene ("Tejon Stage").

A single specimen was collected from locality 4722.

Genus *Cylichnina* Monterosato, 1884

Type species: (by original designation) *Bulla umbilicata* Montague.

Cylichnina tantilla (Anderson and Hanna)

Cylichnella tantilla Anderson and Hanna, 1925:140, pl. 7, figs. 4, 8, 9.
Cylichnina tantilla (Anderson and Hanna). Stewart, 1926:439–441, pl. 27, figs. 2–4. Turner, 1938:67, pl. 20, figs. 9, 10. Vokes, 1939:110, pl. 16, figs. 28, 33, 39.
Hypotypes: UCR 4658/3, locality 4658; UCR 4675/11, locality 4675; UCR 4707/15, locality 4707; UCR 4721/4, locality 4721.
Local occurrence: Turritella uvasana infera, T. uvasana applinae, Ectinochilus supraplicatus, and *E. canalifer* faunas, Juncal and Matilija formations.
Provincial Range: Early Eocene ("Capay Stage") to late Eocene ("Tejon Stage").

Genus *Scaphander* Montfort, 1810

Type species: (by original designation) *Bulla lignaria* Linné.

Subgenus *Mirascapha* Stewart, 1926

Type species: (by original designation) *Cylichna costata* Gabb.

Scaphander (*Mirascapha*) *costatus* (Gabb)

Cylichna costata Gabb, 1864:143, pl. 21, fig. 107.
Scaphander costatus (Gabb). Anderson and Hanna, 1925:139.
Scaphander (*Mirascapha*) *costatus* (Gabb). Stewart, 1926:437, pl. 27, fig. 5. Turner, 1938:67, pl. 17, fig. 16. Vokes, 1939:109, pl. 16, figs. 29, 35.
Scaphander costata (Gabb). Hanna, 1927:329, pl. 57, figs. 2, 3, 5.
Hypotype: UCR 4675/14, locality 4675.

Local occurrence: ***Turritella uvasana applinae*** fauna, Juncal Formation.

Provincial Range: Late Paleocene ("Meganos Stage") to middle Eocene ("Domengine Stage").

A single specimen was collected from locality 4675.

Superfamily PYRAMIDELLACEA

Family PYRAMIDELLIDAE

Genus *Turbonilla* Risso, 1826

Type species: (by original designation) *Turbo lactea* Linné.

Turbonilla gesteri Anderson and Hanna

Turbonilla gesteri Anderson and Hanna, 1925:129, pl. 11, fig. 10.

Hypotype: UCR 4705/9, locality 4705.

Local occurrence: *Ectinochilus supraplicatus* fauna, Juncal Formation.

Provincial Range: Late middle Eocene ("Transition Stage") to late Eocene ("Tejon stage").

Genus *Pyramidella* Lamarck, 1799

Type species: (by original designation) *Trochus dolabrata* Linné.

Pyramidella etheringtoni Hanna

Pyramidella etheringtoni Hanna, 1927:302, pl. 46, figs. 14, 18, 22.

Hypotypes: UCR 4682/17, locality 4682; UCR 4706/21, locality 4706.

Local occurrence: *Turritella uvasana applinae* and *Ectinochilus supraplicatus* faunas, Juncal Formation.

Provincial Range: Middle Eocene ("Domengine Stage") to late middle Eocene ("Transition Stage").

FOSSIL LOCALITIES

UCR 4655 East side of Hot Springs Canyon, 2,240 ft S., 2,000 ft W. of NE. cor. sec. 21, T. 6 N., R. 20 W., Topatopa Mountains quadrangle (1943 ed.). Gray calcareous sandstone bed, 3–5 ft thick, about 150 ft stratigraphically above the base of the Juncal Formation.

UCR 4656 East side of Hot Springs Canyon, 2,440 ft S., 2,075 ft W. of NE. cor. sec. 21, T. 6 N., R. 20 W., Topatopa Mountains quadrangle (1943 ed.). Limestone concretions in dark gray silty mudstone, 10 ft stratigraphically above locality 4655.

UCR 4657 East side of Hot Springs Canyon, 2,650 ft S., 1,900 ft W. of NE. cor. sec. 21, T. 6 N., R. 20 W., Topatopa Mountains quadrangle (1943 ed.). Thin, fine-grained, greenish gray sandstone bed in mudstone facies of the Juncal Formation, 15 ft stratigraphically above locality 4656.

UCR 4658 East side of Hot Springs Canyon, 2,300 ft N., 1,900 ft W., of SE. cor. sec. 21, T. 6 N., R. 20 W., Topatopa Mountains quadrangle (1943 ed.). Thin, fine-grained, greenish gray sandstone bed in mudstone facies, about 340 ft stratigraphically above base of Juncal Formation.

UCR 4659 In bed of Hot Springs Creek, 3,000 ft N., 1,800 ft W. of SE cor. sec. 21, T. 6. N., R. 20 W., Topatopa Mountains quadrangle (1943 ed.). *Turritella* bed in dark gray silty mudstone, about 450 ft stratigraphically above base of Juncal Formation.

UCR 4660 West side of Hot Springs Canyon, 1,650 ft N., 1,650 ft W. of SE. cor. sec. 21, T. 6 N., R. 20 W., Topatopa Mountains quadrangle (1943 ed.). Abundant *Turritella andersoni* in fine-grained greenish gray sandstone bed, 8–12 in thick, in mudstone facies about 500 ft stratigraphically above base of Juncal Formation.

UCR 4661 East side of Hot Springs Canyon, 2,450 ft N., 2,050 ft W. of SE. cor. sec. 21, T. 6 N., R. 20 W., Topatopa Mountains quadrangle (1943 ed.). Abundant *Turritella andersoni* and *T. uvasana infera* in fine-grained greenish gray sandstone bed, 6–12 in thick, in mudstone facies about 305 ft stratigraphically above base of Juncal Formation.

UCR 4662 In bed of Hot Springs Creek, 1,900 ft N., 2,050 ft W. of SE. cor. sec. 21, T. 6 N., R. 20 W., Topatopa Mountains quadrangle (1943 ed.). Abundant *Turritella andersoni* and *T. uvasana infera* in dark gray silty mudstone and in thin silty limestone lenses, about 465 ft stratigraphically above base of Juncal Formation.

UCR 4663 Southwest of Pine Mountain Lodge, 950 ft S., 2,450 ft W. of NE. cor. sec. 14, T. 6 N., R. 22 W., Lion Canyon quadrangle (1943 ed.). Dark brown-weathering limestone concretions in brownish green silty mudstone, about 335 ft stratigraphically above base of Coldwater Sandstone.

UCR 4664 East side of Hot Springs Canyon, 2,200 ft S., 2,650 ft W. of NE. cor. sec. 21, T. 6 N., R. 20 W., Topatopa Mountains quadrangle (1943 ed.). Abundant *Turritella andersoni* in thin, fine-grained, greenish gray sandstone bed, about 175 ft stratigraphically above base of Juncal Formation. Same stratigraphic horizon as locality 4657.

UCR 4665 East side of Hot Springs Canyon, 2,800 ft S., 2,400 ft W. of NE. cor. sec. 21, T. 6 N., R. 20 W., Topatopa Mountains quadrangle (1943 ed.). Limestone concretions in dark gray silty mudstone, about 380 ft stratigraphically above base of Juncal Formation.

UCR 4666 In bed of Hot Springs Creek, 2,200 ft S., 2,900 ft W. of NE. cor. sec. 21, T. 6 N., R. 20 W., Topatopa Mountains quadrangle (1943 ed.). *Turritella andersoni* and other mollusks in fine-grained, greenish gray sandstone bed, about 340 ft stratigraphically above base of Juncal Formation. Approximately same stratigraphic horizon as locality 4658.

UCR 4667 East side of south-draining tributary to Hot Springs Canyon, 1,700 ft S., 1,400 ft E. of NW. cor. sec. 21, T. 6 N., R. 20 W., Topatopa Mountains quadrangle (1943 ed.). Greenish gray sandstone bed, about 175 ft stratigraphically above base of Juncal Formation. Same stratigraphic horizon as localities 4657 and 4664.

UCR 4668 East side of Hot Springs Canyon, 2,500 ft N., 1,400 ft W. of SE. cor. sec. 21, T. 6 N., R. 20 W., Topatopa Mountains quadrangle (1943 ed.). *Velates perversus, Campanilopa dilloni* and other mollusks in gray calcareous sandstone bed, about 150 ft stratigraphically above base of Juncal Formation. Same stratigraphic horizon as locality 4655.

UCR 4670 East side of Hot Springs Canyon, 2,250 ft N., 1,500 ft W. of SE. cor. sec. 21, T. 6 N., R. 20 W., Topatopa Mountains quadrangle (1943 ed.). Thin, fine-grained, greenish gray sandstone bed, about 175 ft stratigraphically above base of Juncal Formation. Same stratigraphic horizon as locality 4657.

UCR 4671 On south side of U.S. Forest Service hiking trail from Scheideck Campground on Reyes Creek to Pine Mountain Lodge, 2,200 ft N., 100 ft W. of SE. cor. sec. 34, T. 7 N., R. 23 W., Reyes Peak quadrangle (1943 ed.). Limestone concretions in light gray silty mudstone, about 100 ft stratigraphically above base of Cozy Dell Shale.

UCR 4672 West of main fork of Piru Creek, 2,450 ft S., 1,300 ft E. of NW. cor. sec. 4, T. 6 N., R. 21 W., Lockwood Valley quadrangle (1943 ed.). Light gray, coarse-grained, calcareous sandstone bed in sandstone facies of Juncal Formation.

UCR 4673 On north side of South Fork of Piru Creek and due north of U.S. Forest Service Campground at Thorn Meadows, 850 ft N., 1,650 ft E. of SW. cor. sec. 4, T. 6 N., R. 21 W., Lockwood Valley quadrangle (1943 ed.). Thin, coarse-grained, pebbly, calcareous sandstone bed in sandstone facies of Juncal Formation.

UCR 4674 On north side of South Fork of Piru Creek, 1,400 ft N., 1,100 ft W. of SE. cor. sec. 5, T. 6 N., R. 21 W., Lockwood Valley quadrangle (1943 ed.). Coarse-grained, conglomeratic, calcareous sandstone bed in sandstone facies of Juncal Formation.

UCR 4675 West of Thorn Meadows Campground on divide between South Fork of Piru Creek and unnamed tributary that passes through Thorn Meadows, 2,000 ft S., 625 ft W. of NE. cor. sec. 8, T. 6 N., R. 21 W., Lockwood Valley quadrangle (1943 ed.). Dark brown-weathering limestone concretions near top of light gray, coarse-grained sandstone bed, 15 ft thick, in siltstone facies of Juncal Formation.

UCR 4676 200 ft east of locality 4675, 1,700 ft S., 500 ft W. of NE. cor. sec. 8, T. 6 N., R. 21 W., Lockwood Valley quadrangle (1943 ed.). Same bed as locality 4675.

UCR 4678 250 ft north of locality 4675, 1,650 ft S., 650 ft W. of NE. cor. sec. 8, T. 6 N., R. 21 W., Lockwood Valley quadrangle (1943 ed.). Same bed as locality 4675.

UCR 4679 At bottom of south-draining gulley west of Thorn Meadows Campground, 1,800 ft N., 2,750 ft E. of SW. cor. sec. 8, T. 6 N., R. 21 W., Lion Canyon quadrangle (1943 ed.). Same bed as locality 4675.

UCR 4680 West of Thorn Meadows Campground, on hillside north of hiking trail from Thorn Meadows to Thorn Point, 1,450 ft N., 2,950 ft E. of SW. cor. sec. 8, T. 6 N., R. 21 W., Lion Canyon quadrangle (1943 ed.). Same bed as locality 4675.

UCR 4681 West of Thorn Meadows Campground, 2,900 ft S., 1,700 ft W. of NE. cor. sec. 8, T. 6 N., R. 21 W., Lockwood Valley quadrangle (1943 ed.). Same bed as locality 4675.

UCR 4682 West of Thorn Meadows Campground, on south side of ridge that forms divide between the South Fork of Piru Creek and the unnamed tributary that flows through Thorn Meadows, 2,200 ft S., 1,600 ft W. of NE. cor. sec. 8, T. 6 N., R. 21 W., Lockwood Valley quadrangle (1943 ed.). Limestone concretions in yellowish gray-weathering sandy siltstone, 15 ft stratigraphically above locality 4675.

UCR 4683 West of Thorn Meadows Campground, along hiking trail from Thorn Meadows to Thorn Point, 1,300 ft N., 2,200 ft E. of SW. cor. sec. 8, T. 6 N., R. 21 W., Lion Canyon quadrangle (1943 ed.). Gray limestone concretions in siltstone facies of Juncal Formation, same stratigraphic horizon as locality 4682.

UCR 4684 150 ft north of locality 4683 on hiking trail from Thorn Meadows to Thorn Point, 1,450 ft N., 2,400 ft E. of SW cor. sec. 8, T. 6 N., R. 21 W., Lion Canyon quadrangle (1943 ed.). Gray limestone concretions in siltstone facies of Juncal Formation, same stratigraphic horizon as localities 4682 and 4683.

UCR 4685 West of Thorn Meadows Campground, on crest of southeast trending spur of ridge that forms the divide between the South Fork of Piru Creek and the unnamed tributary that flows through Thorn Meadows, 2,400 ft S., 1,250 ft W. of NE. cor. sec. 8, T. 6 N., R. 21 W., Lockwood Valley quadrangle (1943 ed.). Gray limestone concretions in siltstone facies of Juncal Formation, 25 ft stratigraphically above locality 4675.

URC 4686 50 ft southwest of locality 4682, 2,200 ft S., 1,650 ft W. of NE. cor. sec. 8, T. 6 N., R. 21 W., Lockwood Valley quadrangle (1943 ed.). Gray limestone concretions in siltstone facies of Juncal Formation, same stratigraphic horizon as locality 4685.

UCR 4687 West of Thorn Meadows Campground, on east side of south-trending spur of ridge that forms the divide between the South Fork of Piru Creek and the unnamed tributary that flows through Thorn Meadows, 1,750 ft N., 3,150 ft E. of SW. cor. sec. 8, T. 6 N., R. 21 W., Lion Canyon quadrangle (1943 ed.). Gray limestone concretions in siltstone facies of Juncal Formation, same stratigraphic horizon as localities 4685 and 4686.

UCR 4688 West of Thorn Meadows Campground, 500 ft N. 45 W. of locality 4687, 2,400 ft N., 2,750 ft E. of SW. cor. sec. 8, T. 6 N., R. 21 W., Lion Canyon quadrangle (1943 ed.). Fossils collected from siltstone and from limestone concretions in the siltstone, 75 ft stratigraphically above locality 4675.

UCR 4690 West of Thorn Meadows Campground, on crest of ridge that forms divide between the South Fork of Piru Creek and the unnamed tributary that flows through Thorn Meadows, 2,600 ft S., 2,700 ft E. of NW. cor. sec. 8, T. 6 N., R. 21 W., San Guillermo quadrangle (1943 ed.). *Turritella scrippsensis* and other mollusks in hard, ledge-forming concretionary, calcareous sandstone bed, 95 ft stratigraphically above locality 4675. Siltstone facies of Juncal Formation.

UCR 4694 On crest of divide between the Piru Creek and Alamo Creek watersheds, 650 ft S., 1,600 ft W. of NE. cor. sec. 35, T. 7 N., R. 22 W., San Guillermo quadrangle (1943 ed.). Abundant *Turritella andersoni lawsoni* and *T. uvasana applinae* in thin, fine-grained, calcareous sandstone bed in siltstone facies of Juncal Formation.

UCR 4695 About 100 ft east of U.S. Forest Service hiking trail from Fishbowls Campground on Piru Creek to Pine Mountain Lodge, 2,500 ft N., 500 feet E. of SW. cor. sec. 6, T. 6 N., R. 21 W., San Guillermo quadrangle (1943 ed.). Oyster biostrome, about 2 ft thick, in lens of greenish brown silty mudstone within the sandstone facies of the Juncal Formation.

UCR 4696 About 4,000 ft SW. of locality 4695, on U.S. Forest Service hiking trail from Fishbowls Campground to Pine Mountain Lodge, 1,300 ft S., 400 ft W. of NE. cor. sec. 12, T. 6 N., R. 22 W., San Guillermo quadrangle (1943 ed.). Oysters and other mollusks in hard, dark brown-weathering, calcareous sandstone concretions in uppermost tongue of siltstone facies of Juncal Formation.

UCR 4697 1,300 ft due north of Thorn Point, on slope below massive, ledge-forming, 60-ft thick bed of light gray conglomeratic sandstone, 1,750 ft N., 3,000 ft E. of SW. cor. sec. 17, T. 6 N., R. 21 W., Lion Canyon quadrangle (1943 ed.). Collection from dark brown–weathering calcareous sandstone concretions near top of uppermost tongue of siltstone facies of Juncal Formation.

UCR 4698 900 ft north of Thorn Point Fire Lookout Tower and just south of prominent ledge-forming sandstone bed mentioned above, 1,400 ft N., 2,500 ft E. of SW cor. sec. 17, T. 6 N., R. 21 W., Lion Canyon quadrangle (1943 ed.). Oyster biostrome in green silty mudstone bed, about 5 ft thick, in sandstone facies of Juncal Formation.

UCR 4699 On crest of northwest-trending ridge south of the main fork of Park Canyon, 1,600 ft S., 2,600 ft E. of NW. cor. sec. 13, T. 7 N., R. 22 W., San Guillermo quadrangle (1943 ed.). Grayish orange–weathering, calcareous, pebbly sandstone bed, about 2 ft thick, in siltstone facies of Juncal Formation.

UCR 4700 Bottom of north-draining tributary to main fork of Park Canyon, 2,200 ft S., 2,400 ft E. of NW. cor. sec. 13, T. 7 N., R. 22 W., San Guillermo quadrangle (1943 ed.). Abundant *Turritella andersoni lawsoni* var. *secondaria* and other mollusks in thin, grayish orange–weathering, concretionary, calcareous sandstone bed in siltstone facies of Juncal Formation.

UCR 4701 In bottom of gully at head of long, north-draining, tributary to Wagon Road Canyon, 1800 ft S., 1,600 ft W. of NE. cor. sec. 22, T. 7 N., R. 22 W., San Guillermo quadrangle (1943 ed.). Abundant *Turritella andersoni lawsoni* in thin, fine-grained, calcareous sandstone bed in siltstone facies of Juncal Formation.

UCR 4702 On crest of ridge 300 ft SE. of locality 4701, 2,000 ft S., 1,400 ft W. of NE. cor. sec. 22, T. 7 N., R. 22 W., San Guillermo quadrangle (1943 ed.). Abundant *Turritella uvasana applinae* in thin, fine-grained, calcareous sandstone bed in siltstone facies of Juncal Formation.

UCR 4703 At elevation of about 5,250 ft on hillside south of south fork of Park Canyon, 2,400 ft S. of NE. cor. sec. 23, T. 7 N., R. 22 W., San Guillermo quadrangle (1943 ed.). From thin, fine-grained, concretionary, calcareous sandstone bed in siltstone facies of Juncal Formation.

UCR 4705 On north side of first west-draining tributary to Alamo Creek south of the mouth of Wagon Road Canyon, 400 ft N., 800 ft W. of SE. cor. sec. 21, T. 7 N., R. 22 W., San Guillermo quadrangle (1943 ed.). From thin, gray, concretionary, calcareous sandstone bed in uppermost tongue of siltstone facies of Juncal Formation.

UCR 4706 West of Reyes Peak, in road cut on U.S. Forest Service Pine Mountain truck trail, 650 ft N., 1,900 ft E. of SW cor. sec. 2, T. 6 N., R. 23 W., Reyes Peak quadrangle (1943 ed.). From dark brown–weathering calcareous sandstone concretions in uppermost tongue of siltstone facies of Juncal Formation.

UCR 4707 On south side of small ridge south of the second major west-draining tributary to Alamo Creek above the mouth of Wagon Road Canyon, 1,700 ft N., 1,400 ft W. of SE. cor. sec. 27, T. 7 N., R. 22 W., San Guillermo quadrangle (1943 ed.). From dark brown–weathering calcareous sandstone concretions in uppermost tongue of siltstone facies of Juncal Formation.

UCR 4708 600 ft S. 65 E. of locality 4707, 1,500 ft N., 800 ft W. of SE. cor. sec. 27, T. 7 N., R. 22 W., San Guillermo quadrangle (1943 ed.). Fossils collected from dark brown–weathering calcareous sandstone concretions in uppermost tongue of siltstone facies of Juncal Formation.

UCR 4714 On ridge northeast of Pine Mountain Lodge, 1250 ft N., 2,600 ft E. of SW. cor. sec. 12, T. 6 N., R. 22 W., Lion Canyon quadrangle (1943 ed.). From dark brown–weathering calcareous sandstone concretions, about 75 ft stratigraphically above base of Matilija Sandstone.

UCR 4715 90 ft south of Elevation 6679 on ridge northeast of Pine Mountain Lodge, 850 ft N., 2,200 ft W. of SE. cor. sec. 12, T. 6 N., R. 22 W., Lion Canyon quadrangle (1943 ed.). From dark brown–weathering calcareous sandstone concretions, about 110 ft stratigraphically above base of Matilija Sandstone.

UCR 4716 500 ft N. 65 W. of Elevation 6679 on ridge northeast of Pine Mountain Lodge, 1,050 ft N., 2,500 ft E. of SW. cor. sec. 12, T. 6 N., R. 22 W., Lion Canyon quadrangle (1943 ed.). From dark brown–weathering calcareous sandstone concretions, about 200 ft stratigraphically above base of Matilija Sandstone.

UCR 4717 1,000 ft S. 80 E. of Elevation 6412 on ridge northeast of Pine Mountain Lodge, 200 ft N., 1,300 ft E. of SW. cor. sec. 12, T. 6 N., R. 22 W., Lion Canyon quadrangle (1943 ed.). From dark brown–weathering calcareous sandstone concretions in tongue of sandstone facies of Juncal Formation, about 650 ft stratigraphically above base of Matilija Sandstone.

UCR 4718 1,600 ft S. 80 E. of Elevation 6412 on ridge northeast of Pine Mountain Lodge, 200 ft N., 1,750 ft E. of SW. cor. sec. 12, T. 6 N., R. 22 W., Lion Canyon quadrangle. Abundant *Turritella uvasana* s.s. in dark brown–weathering calcareous sandstone concretions in tongue of sandstone facies of Juncal Formation, about 650 ft stratigraphically above base of Matilija Sandstone. Same stratigraphic horizon at locality 4717.

UCR 4719 1,600 ft S. 70 E. of Elevation 6412 on ridge northeast of Pine Mountain Lodge, 200 ft S., 1,600 ft E. of NW. cor. sec. 13, T. 6 N., R. 22 W., Lion Canyon quadrangle (1943 ed.). Fossils collected from dark brown–weathering calcareous sandstone concretions, about 800 ft stratigraphically above base of Matilija Sandstone.

UCR 4720 1,600 ft S. 68 E. of Elevation 6412 on ridge northeast of Pine Mountain Lodge, 400 ft S., 1,450 ft E. of NW. cor. sec. 13, T. 6 N., R. 22 W., Lion Canyon quadrangle (1943 ed.). From dark brown–weathering calcareous sandstone concretions, about 840 ft stratigraphically above base of Matilija Sandstone.

UCR 4721 250 ft S. 45 W. of Elevation 6413 on ridge east of Pine Mountain Lodge, 1,200 ft S., 2,550 ft E. of NW. cor. sec. 13, T. 6 N., R. 22 W., Lion Canyon quadrangle (1943 ed.). *Turritella uvasana sargeanti* and other mollusks collected from dark brown–weathering calcareous sandstone concretions, about 1,030 ft stratigraphically above base of Matilija Sandstone.

UCR 4722 500 ft N. 25 E. of Elevation 6412 on ridge northeast of Pine Mountain Lodge, 900 ft N., 400 ft E. of SW. cor. sec. 12, T. 6 N., R. 22 W., Lion Canyon quadrangle (1943 ed.). From dark brown–weathering calcareous sandstone concretions in tongue of sandstone facies of Juncal Formation, about 650 ft stratigraphically above base of Matilija Sandstone. Same stratigraphic horizon as localities 4717 and 4718.

UCR 4723 200 ft S. 40 W. of Elevation 6679 on ridge northeast of Pine Mountain Lodge, 750 ft N., 2,250 ft W. of SE. cor. sec. 12, T. 6 N., R. 22 W., Lion Canyon quadrangle (1943 ed.). Abundant *Turritella uvasana* s.s. in thin, concretionary, calcareous sandstone bed, about 230 ft stratigraphically above base of Matilija Sandstone.

UCR 4724 25 ft north of Elevation 6413 on ridge east of Pine Mountain Lodge, 900 ft S., 2,600 ft E. of NW. cor. sec. 13, T. 6 N., R. 22 W., Lion Canyon quadrangle (1943 ed.). Oyster biostrome, about 980 ft stratigraphically above base of Matilija Sandstone.

UCR 4726 1,800 ft almost due east of Pine Mountain Lodge, 650 ft S., 700 ft E. of NW. cor. sec. 13, T. 6 N., R. 22 W., Lion Canyon quadrangle (1943 ed.). *Turritella uvasana sargeanti* and other mollusks collected from dark brown–weathering calcareous sandstone concretions, about 1,030 ft stratigraphically above base of Matilija Sandstone. Same stratigraphic horizon as locality 4721.

UCR 4727 800 ft northwest of Pine Mountain Lodge, on U.S. Forest Service hiking trail, 1,600 ft W. of NE. cor. sec. 14, T. 6 N., R. 22 W., Lion Canyon quadrangle (1943 ed.). From dark gray, fine-grained, limestone concretions in silty mudstone, about 165 ft stratigraphically below top of Matilija Sandstone.

UCR 4728 400 ft W. of locality 4727 and 2,000 ft W. of NE. cor. sec. 14, T. 6 N., R. 22 W., Lion Canyon quadrangle (1943 ed.). From dark gray, fine-grained, limestone concretions in silty mudstone, about 60 ft stratigraphically below top of Matilija Sandstone.

UCR 4729 700 ft S. 55 E. of Pine Mountain Lodge, 900 ft S., 500 ft W. of NE. cor. sec. 14, T. 6 N., R. 22 W., Lion Canyon quadrangle (1943 ed.). From dark gray, fine-grained, limestone concretions in silty mudstone, about 20 ft stratigraphically below top of Matilija Sandstone.

UCR 4730 2,000 ft N. 80 W. of Pine Mountain Lodge, 350 ft S., 2,250 ft E. of NW. cor. sec. 14, T. 6 N., R. 22 W., Lion Canyon quadrangle (1943 ed.). *Venericardia* biostrome in light gray, medium-grained, calcareous sandstone, about 280 ft stratigraphically above base of Coldwater Sandstone.

UCR 4731 1,100 ft S. 75 E. of Pine Mountain Lodge, 950 ft S., 100 ft W. of NE. cor. sec. 14, T. 6 N., R. 22 W., Lion Canyon quadrangle (1943 ed.). From dark gray, fine-grained, limestone concretions in silty mudstone, about 95 ft stratigraphically below top of Matilija Sandstone.

UCR 4732 250 ft S. 55 E. of Pine Mountain Lodge, 750 ft S., 1050 ft W. of NE. cor. sec. 14, T. 6 N., R. 22 W., Lion Canyon quadrangle (1943 ed.). From dark gray, fine-grained, limestone concretions in silty mudstone, about 80 ft stratigraphically below top of Matilija Sandstone.

UCR 4738 Near base of hill on west side of Beartrap Creek, about 2,000 ft upstream from its confluence with Alamo Creek, 1,300 ft N., 450 ft E. of SW. cor. sec. 23, T. 7 N., R. 23 W., Reyes Peak quadrangle (1943 ed.). From dark brown–weathering calcareous sandstone concretions, about 200 ft stratigraphically above base of Matilija Sandstone.

UCR 4739 650 ft N. 75 E. of Elevation 4395, 100 ft S., 400 ft E. of NW. cor. sec. 26, T. 7 N., R. 23 W., Reyes Peak quadrangle (1943 ed.). From fine-grained brown sandstone, about 500 ft stratigraphically above base of Matilija Sandstone.

UCR 4740 100 ft S. 45 W. of Elevation 4395, 300 ft S., 300 ft W. of NE. cor. sec. 27, T. 7 N., R. 23 W., Reyes Peak quadrangle (1943 ed.). *Turritella* cf. *T. uvasana* s.s. collected from gray limestone concretions, about 150 ft stratigraphically above base of Cozy Dell Shale.

UCR 4741 East of confluence of Alamo Creek, Dry Canyon, and Cuyama River valley, 800 ft S., 2,700 ft W. of NE. cor. sec. 23, T. 7 N., R. 23 W., Reyes Peak quadrangle (1943 ed.). From dark brown–weathering calcareous sandstone concretions in lower part of Matilija Sandstone.

UCR 4743 2,500 ft N. 80 E. of Elevation 4846 in center of enclosed depression on plateau southeast of Wegis Ranch, 2,400 ft S., 1,400 ft E. of NW. cor. sec. 29, T. 7 N., R. 22 W., San Guillermo quadrangle (1943 ed.). From dark brown–weathering calcareous sandstone concretions, about 200 ft stratigraphically above base of Matilija Sandstone.

UCR 4744 125 ft S. 65 W. of Elevation 5194 on plateau southeast of Wegis Ranch, 1,700 ft N., 2,200 ft E. of SW. cor. sec. 29, T. 7 N., R. 22 W., San Guillermo quadrangle (1943 ed.). From dark brown–weathering calcareous sandstone concretions, about 100 ft stratigraphically above base of Matilija Sandstone.

UCR 4745 On east side of hill east of confluence of Alamo Creek, Dry Canyon, and the Cuyama River valley, 400 ft S., 1,400 ft W. of NE. cor. sec. 23, T. 7 N., R. 23 W., Reyes Peak quadrangle (1943 ed.). *Venericardia hornii* s.s. collected from dark brown–weathering calcareous sandstone concretions in lower part of Matilija Sandstone.

UCR 4746 300 ft S. 50 W. of Elevation 4072 on ridge south of the mouth of Beartrap Creek, 1,750 ft N., 1,200 ft E. of SW. cor. sec. 23, T. 7 N., R. 23 W., Reyes Peak quadrangle (1943 ed.). *Venericardia hornii* s.s. and other mollusks collected from dark brown–weathering calcareous sandstone concretions, about 150 ft stratigraphically above base of Matilija Sandstone.

UCR 4747 Just east of Elevation 4072 on ridge south of mouth of Beartrap Creek, 1,950 ft N., 1,500 ft E. of SW. cor. sec. 23, T. 7 N., R. 23 W., Reyes Peak quadrangle (1943 ed.). From dark brown–weathering calcareous sandstone concretions, about 50 ft stratigraphically above base of Matilija Sandstone.

UCR 4750 On crest of ridge 2,700 ft W. of Elevation 6602 on San Guillermo Mountain, 500 ft N., 700 ft W. of SE. cor. sec. 13, T. 7 N., R. 22 W., San Guillermo quadrangle (1943 ed.). From dark brown–weathering, calcareous, conglomeratic sandstone lens in siltstone facies of Juncal Formation.

UCR 4751 On crest of southwest-trending ridge northeast of the main fork of Piru Creek, 1,350 ft S., 3,050 ft E. of NW. cor. sec. 32, T. 7 N., R. 21 W., San Guillermo quadrangle (1943 ed.). *Claibornites diegoensis* collected from yellowish orange–weathering calcareous sandstone lens in siltstone facies of Juncal Formation.

UCR 4752 600 ft S. 45 W. of locality 4751, 1,750 ft S., 2,450 ft E. of NW. cor. sec. 32, T. 7 N., R. 21 W., San Guillermo quadrangle (1943 ed.). *Isognomon* n. sp.? and other mollusks collected from lens of calcareous sandy conglomerate in sandstone facies of Juncal Formation.

UCR 4753 300 ft due east of locality 4708, 1,400 ft N., 400 ft W. of SE. cor. sec. 27, T. 7 N., R. 22 W., San Guillermo quadrangle (1943 ed.). From dark brown–weathering calcareous sandstone concretions in uppermost tongue of siltstone facies of Juncal Formation.

LITERATURE CITED

ABBOTT, R. T.
1954. American seashells. N.Y.: D. Van Nostrand Co. 541 pp., 40 pls.
1958. The marine mollusks of Grand Cayman Island, British West Indies. Monogr. of the Acad. of Natur. Sci. of Phila., no. 11. 138 pp., 5 pls.
1959. The helmet shells of the world (Cassidae), Part 1. Indo-Pacific Mollusca, 2(9): 15–200, pls. 9–187.

ALLEN, J. R. L.
1965. A review of the origin and characteristics of Recent alluvial sediments. Sedimentology, 5:89–191.

AMERICAN COMMISSION ON STRATIGRAPHIC NOMENCLATURE
1970. Code of stratigraphic nomenclature. Tulsa: American Association of Petroleum Geologists. 22 pp.

ANDERSON, F. M., and G. D. HANNA
1925. Fauna and stratigraphic relations of the Tejon Eocene at the type locality in Kern County California. Calif. Acad. Sci. Occas. Papers, no. 11. 249 pp., 16 pls.

ARNOLD, R.
1909. Paleontology of the Coalinga district, Fresno and Kings counties, California. U.S. Geol. Surv. Bull. 396. 173 pp., 30 pls.

ARNOLD, R., and R. V. ANDERSON
1910. Geology and oil resources of the Coalinga district, California. U.S. Geol. Surv. Bull. 398. 354 pp., 52 pls.

BAILEY, T. L., and R. H. JAHNS
1954. Geology of the Transverse Range province, southern California. Calif. Div. Mines Bull. 170. Pp. 83–106.

BARRELL, J.
1925. Marine and terrestrial conglomerates. Geol. Soc. Amer. Bull., 36:279–342.

BENTSON, H.
1940. A systematic study of the fossil gastropod *Exilia*. Univ. Calif. Publ. Bull. Dept. Geol. Sci., 25(5):199–238, pls. 1–3.

BRABB, E. E., BUKRY, D., and R. L. PIERCE
1971. Eocene (Refugian) nannoplankton in the Church Creek Formation near Monterey, central California. U.S. Geol. Surv. Prof. Paper 750-C. Pp. 44–47.

CARMAN, M. F., JR.
1964. Geology of the Lockwood Valley area, Kern and Ventura counties, California. Calif. Div. Mines and Geol. Spec. Rep. 81. 62 pp., 5 pls.

CERNOHORSKY, W. O.
1970. Systematics of the families Mitridae and Volutomitridae (Mollusca: Gastropoda). Auckland Institute and Museum Bull. 8. 190 pp., 18 pls., 222 text figs., 3 tables.

CLARK, B. L.
1918. The San Lorenzo series of middle California. Univ. Calif. Publ. Bull. Dept. Geol. Sci., 11(2): 45–234, pls. 3–24.
1925. Pelecypoda from the marine Oligocene of western North America. Univ. Calif. Publ. Bull. Dept. Geol. Sci., 15:69–136, pls. 8–22.
1926. The Domengine Horizon, middle Eocene of California. Univ. Calif. Publ. Bull. Dept. Geol. Sci., 16:99–118.
1934. A new genus and two new species of Lamellibranchiata from the middle Eocene of California. J. Paleontol., 8:270–272, pl. 37.
1935. Tectonics of the Mount Diablo and Coalinga areas, middle Coast Ranges of California. Geol. Soc. Amer. Bull., 46:1025–1078, pls. 88–90.
1938. Fauna from the Markley Formation (upper Eocene) on Pleasant Creek, California. Geol. Soc. Amer. Bull., 49:683–730, 4 pls.
1942. New middle Eocene gastropods from California. J. Paleontol. 16(1):116–119, pl. 19.
1943. Notes on California Tertiary correlation. Calif. Div. Mines Bull. 118. Pp. 187–191.

CLARK, B. L., and D. K. PALMER
1923. Revision of the *Rimella*-like gastropods from the West Coast of North America. Univ. Calif. Publ. Bull. Dept. Geol. Sci., 14:277–288, pl. 51.

CLARK, B. L., and H. E. VOKES
1936. Summary of the marine Eocene sequence of western North America. Geol. Soc. Amer. Bull., 47:851–878, pls. 1, 2.

CLARK, B. L., and A. O. WOODFORD
1927. The geology and paleontology of the type section of the Meganos Formation (lower middle Eocene) of California. Univ. Calif. Publ. Bull. Dept. Geol. Sci., 17(2):63–142, pls. 14–22.

COLEMAN, J. M., and S. M. GAGLIANO
1965. Sedimentary structures: Mississippi River deltaic plain. *In* G. V. Middleton, ed., Primary sedimentary structures and their hydrodynamic interpretation. Soc. Econ. Paleontol. and Mineral. Spec. Publ. 12. Pp. 133–148.

CONRAD, T. A.
1855. Report on the fossil shells collected in California by W. P. Blake. U.S. 33d. Congr., 1st Sess., House Ex. Doc. no. 129. Pp. 5–20.
1857. Descriptions of fossil shells. Pacific Railroad Rep. 5. Pp. 317–329, pls. 2–9.

COOPER, J. G.
1894. Catalogue of California fossils, parts 2–5. Calif. State Mining Bur. Bull. 4. 65 pp., 6 pls.

COSSMANN, A. E. M.
1913. Rectifications de nomenclature. Rev. Crit. Paleoz., 17:61–64.

COX, L. R., et al.
1969. Mollusca 6 (part N), vols. 1 and 2. *In* R. C. Moore, ed., Treatise on invertebrate paleontology. Lawrence: University of Kansas Press, Pp. 1–952.

CROOK, T. H., and J. M. KIRBY
1935. The Capay Formation. Geol. Soc. Amer. Proc. for 1934 (Abstr.). Pp. 334–335.

CROWELL, J. C., and T. SUSUKI
1959. Eocene stratigraphy and paleontology, Orocopia Mountains, southeastern California. Geol. Soc. Amer. Bull., 70:581–592, 3 pls.

DALL, W. H.
1909. Contributions to the Tertiary paleontology of the Pacific Coast. I. The Miocene of Astoria and Coos Bay, Oregon. U.S. Geol. Surv. Prof. Paper 59. 278 pp., 23 pls.

DIBBLEE, T. W., JR.
1950. Geology of southwestern Santa Barbara County, California. Calif. Div. Mines Bull. 150. 95 pp., 17 pls.
1966. Geology of the central Santa Ynez Mountains, Santa Barbara County, California. Calif. Div. Mines and Geol. Bull. 186. 99 pp., 4 pls.

DICKERSON, R. E.
1913. Fauna of the Eocene at Marysville Buttes, California. Univ. Calif. Publ. Bull. Dept. Geol. Sci., 7(12):257–298, pls. 11–14.
1914. The fauna of the *Siphonalia sutterensis* zone in the Roseburg quadrangle, Oregon. Calif. Acad. Sci. Proc., 4th ser., 4:113–128, pls. 11, 12.
1915. Fauna of the type Tejon: Its relation to the Cowlitz phase of the Tejon group of Washington. Calif. Acad. Sci. Proc., 4th ser., 5(3):33–98, pls. 1–11.
1916. Stratigraphy and fauna of the Tejon Eocene of California. Univ. Calif. Publ. Bull. Dept. Geol. Sci., 9(17):363–524, pls. 36–46.
1917. Climate and its influence upon the Oligocene faunas of the Pacific Coast, with descriptions of some new species from the *Molopophorus lincolnensis* zone. Calif. Acad. Sci. Proc., 4th ser., 7(6):157–192, pls. 27–31.

DICKINSON, W. R.
1969. Geologic problems in the mountains between Ventura and Cuyama. *In* Upper Sespe Creek: Pacific Section, Soc. Econ. Paleontologists and Mineralogists, Field Trip Guidebook. Pp. 1–23.

DOEGLAS, D. J.
1962. The structure of sedimentary deposits of braided rivers. Sedimentology, 1:167–190.

Dreyer, F. E.
1935. Geology of a portion of Mt. Pinos quadrangle, Ventura County, California. Unpub. M.A. thesis, Dept. of Geology, University of California, Los Angeles.

Durham, J. W.
1937. Gastropods of the family Epitoniidae from Mesozoic and Cenozoic rocks of the west coast of North America, including one new species by F. E. Turner and one by R. A. Bramkamp. J. Paleontol. 11 (6) :479–512, pls. 56, 57.
1942. Notes on Pacific Coast Galeodeas. J. Paleontol. 16 (2) :183–191, pls. 29, 30.

Effinger, W. L.
1938. The Gries Ranch fauna (Oligocene) of western Washington. J. Paleontol. 12 (4) :355–390, pls. 45–47.

Gabb, W. M.
1864. Description of the Cretaceous fossils. Geological Survey of California: Paleontology, vol. 1, section 4. Pp. 55–236, pls. 9–32.
1868. An attempt at a revision of the two families Strombidae and Aporrhaidae. Amer. J. Conchology, 4:137–149, pls. 13, 14.
1869. Cretaceous and Tertiary fossils. Geological Survey of California: Palaeontology, vol. 2, 299 p., 36 pls.

Gardner, J. A., and E. Bowles
1934. Early Tertiary species of gastropods from the Isthmus of Tehuantepec. J. Wash. Acad. Sci., 24:241–248.

Gazin, C. L.
1930. Geology of the central portion of the Mt. Pinos quadrangle, Ventura and Kern counties, southern California. Unpub. Ph.D. dissertation, California Institute of Technology, Pasadena.

Hanna, G D.
1924. Rectifications of nomenclature. Calif. Acad. Sci. Proc., 4th ser., 13:151–186.

Hanna, G D., and L. G. Hertlein
1943. Characteristic fossils of California. Calif. Div. Mines Bull. 118, part 2:165–182, figs. 60–67.
1949. Two new species of gastropods from the middle Eocene of California. J. Paleontology, 23 (4) :392–394, pl. 77.

Hanna, M. A.
1925. Notes on the genus *Venericardia* from the Eocene of the west coast of North America. Univ. Calif. Publ. Bull. Dept. Geol. Sci., 15 (8) :281–306, pls. 36–44.
1926. Geology of the La Jolla quadrangle, California. Univ. Calif. Publ. Bull. Dept. Geol. Sci., 16 (7) :187–246, pls. 17–23, 1 map.
1927. An Eocene invertebrate fauna from the La Jolla quadrangle, California. Univ. Calif. Publ. Bull. Dept. Geol. Sci., 16 (8) :247–398, pls. 24–57.

Hartman, D. C.
1957. Geology of the upper Wagon Road Canyon area, southern California. Unpub. M.A. thesis, University of California, Los Angeles.

Hedgepeth, J. W.
1953. An introduction to the zoogeography of the northwestern Gulf of Mexico with reference to the invertebrate fauna. Univ. Texas Inst. Marine Sci. Publ., 3 (1) :111–223.

Hickman, C. J. S.
1969. The Oligocene marine molluscan fauna of the Eugene Formation in Oregon. Univ. Oregon Museum Nat. Hist. Bull. 16. 112 pp., 14 pls.

Hill, M. L.
1954. Tectonics of faulting in southern California. Calif. Div. Mines Bull. 170. Pp. 5–13.

Hill, M. L., and T. W. Dibblee, Jr.
1953. San Andreas, Garlock and Big Pine faults, California. Geol. Soc. Amer. Bull., 64:443–458.

Hoyt, J. H., and R. J. Weimer
1965. The origin and significance of *Ophiomorpha* (*Halymenites*) in the Cretaceous of the Western Interior. 19th Field Conf. Wyo. Geol. Assoc. Guidebook. Pp. 203–207, figs. 1–7.

JAMES, G. T.

1963. Paleontology and nonmarine stratigraphy of the Cuyama Valley badlands, California: Part 1. Geology, faunal interpretations, and systematic descriptions of Chiroptera, Insectivora and Rodentia. Univ. Calif. Publ. Geol. Sci., vol. 45, 178 pp., 8 pls.

JENNINGS, C. W., and R. G. STRAND

1969. Geologic map of California, Los Angeles sheet (Olaf P. Jenkins edition). San Francisco: California Division of Mines and Geology.

JESTES, E. C.

1963. A stratigraphic study of some Eocene sandstone, northeastern Ventura Basin, California. Unpub. Ph.D. dissertation, University of California, Los Angeles.

KEEN, A. M.

1944. Catalogue and revision of the gastropod subfamily Typhinae. J. Paleontol., 18(1):50–72, 20 text fig.

1958. Sea shells of tropical west America, marine mollusks from Lower California to Colombia. Stanford: Stanford University Press. 624 pp.

1963. Marine molluscan genera of western North America. Stanford: Stanford University Press. 126 pp., illus.

KEEN, A. M., and H. BENTSON

1944. Checklist of California Tertiary marine Mollusca. Geol. Soc. Amer. Spec. Paper 56. 280 pp.

KELLEY, F. R.

1943. Eocene stratigraphy in western Santa Ynez Mountains, Santa Barbara County, California. Amer. Assoc. Petrol. Geol. Bull., 27(1):1–19.

KENNEDY, M. P., and G. W. MOORE

1971. Stratigraphic relations of upper Cretaceous and Eocene formations, San Diego coastal area, California. Amer. Assoc. Petrol. Geol. Bull., 55(5):709–722.

KERR, P. F., and H. G. SCHENCK

1928. Significance of the Matilija Overturn (Santa Ynez Mountains, California). Geol. Soc. Amer. Bull., 39(4):1087–1102.

KEW, W. S. W.

1924. Geology and oil resources of a part of Los Angeles and Ventura Counties, California. U.S. Geol. Surv. Bull. 753. 202 pp.

KIESSLING, E. W.

1958. Geology of the southwest portion of the Lockwood Valley quadrangle, Ventura County, California. Unpub M.A. thesis, University of California, Los Angeles.

KLEINPELL, R. M., and D. W. WEAVER

1963. Oligocene biostratigraphy of the Santa Barbara Embayment, California. Univ. Calif. Publ. Geol. Sci., vol. 43. 249 pp., 38 pls., 8 figs.

LADD, H. S.

1951. Brackish-water and marine assemblages of the Texas Coast, with special reference to mollusks. Univ. Texas Inst. Marine Sci. Publ., 2(1):129–163.

LIPPS, J. H.

1967. Planktonic foraminifera, intercontinental correlation and age of mid-Cenozoic microfaunal stages. J. Paleontol., 41(4):994–999.

MACGINITIE, G. E., and N. MACGINITIE

1949. Natural history of marine animals. New York: McGraw-Hill Book Co. 473 pp., illus.

MALLORY, V. S.

1959. Lower Tertiary biostratigraphy of the California Coast Ranges. Tulsa: American Association of Petroleum Geologists Publication. 416 pp., illus.

MARKS, J. G.

1941. Stratigraphy of the Tejon Formation in its type area, Kern County, California. Unpub. M.A. thesis, Stanford University.

1943. Type locality of the Tejon Formation. Calif. Div. Mines Bull. 118. Part 3, pp. 534–538.

MERRIAM, C. W.

1941. Fossil Turritellas from the Pacific Coast region of North America. Univ. Calif. Publ. Bull. Dept. Geol. Sci., vol. 26. Pp. 1–214, pls. 1–41.

MERRIAM, C. W., and F. E. TURNER
1937. The Capay middle Eocene of northern California. Univ. Calif. Publ. Bull. Dept. Geol. Sci., 24(6):91–114, pls. 5, 6.
MERRILL, W. R.
1954. Geology of the Sespe Creek–Pine Mountain area, Ventura County [California]. Calif. Div. Mines Bull. 170. Map sheet 3.
NEWMAN, P. V.
1959. Geology of the Round Spring Canyon area, northwestern Ventura County, California. Unpub. M.A. thesis, University of California, Los Angeles.
PACKARD, E. L.
1916. Mesozoic and Cenozoic Mactrinae of the Pacific Coast of North America. Univ. Calif. Publ. Bull. Dept. Geol. Sci., 9(15):261–360, pls. 12–35.
PAGE, B. M., MARKS, J. G., and G. W. WALKER
1951. Stratigraphy and structure of the mountains northeast of Santa Barbara, California. Amer. Assoc. Petrol. Geol. Bull., 35(8):1727–1780.
PALMER, K. V.
1937. The Claibornian Scaphopoda, Gastropoda, and dibranchiate Cephalopoda of the southeastern United States. Bull. Amer. Paleont., 7(32), parts 1 and 2. 730 pp., 90 pls.
PARKER, R. H.
1959. Macro-invertebrate assemblages of central Texas coastal bays and Laguna Madre. Amer. Assoc. Petrol. Geol. Bull., 43(9):2100–2166, 32 figs., 6 pls.
1964. Zoogeography and ecology of macro-invertebrates of Gulf of California and continental slope of western Mexico. *In* T. H. van Andel and G. G. Shor, Jr., eds., Geology of the Gulf of California. Amer. Assoc. Petrol. Geol. Mem. 3. Pp. 331–376, 10 pls., 21 text figs., 4 tables.
PETTIJOHN, F. J.
1957. Sedimentary rocks. New York: Harper and Brothers. 718 pp., illus.
POYNER, W. D.
1960. Geology of the San Guillermo area and its regional correlation, Ventura County, California. Unpub. M.A. thesis, University of California, Los Angeles.
1965. Relationship of the Big Pine, San Guillermo, and Ozena faults, northwestern Ventura County, California [abstr.]. Amer. Assoc. Petrol. Geol. Bull., 49(7):1088.
REDWINE, L. E., et al.
1952. Cenozoic correlation section, paralleling north and south margins, western Ventura basin, from Point Conception to Ventura and Channel Islands, California. Amer. Assoc. Petrol. Geol. Pacific Sec., Geol. Names and Correlations Comm., Subcomm. on Cenozoic. 2 sheets.
REINHART, P. W.
1943. Mesozoic and Cenozoic Arcidae from the Pacific Slope of North America. Geol. Soc. Amer. Spec. Paper 47. 117 pp., 15 pls., 3 figs.
RICKETTS, E. F., and J. CALVIN
1960. Between Pacific tides. Stanford: Stanford University Press. 502 pp., illus.
RUTH, J. W.
1942. The molluscan genus *Siphonalia* of the Pacific Coast Tertiary. Univ. Calif. Publ. Bull. Dept. Geol. Sci., 26(3):287–306, pls. 47, 48.
SCHENCK, H. G.
1926. Cassididae of western North America. Univ. Calif. Publ. Bull. Dept. Geol. Sci., 16:69–98, pls. 12–15.
1936. Nuculid bivalves of the genus *Acila*. Geol. Soc. Amer. Spec. Paper 4. 149 pp., 18 pls., 15 figs.
SCHLEE, J. S.
1952. Geology of the Mutau Flat area, Ventura County, California. Unpub. M.A. thesis, University of California, Los Angeles.
SCHMIDT, R. R.
1971. Planktonic foraminiferal zonation from the lower Tertiary of California: Paleogeographic and paleoclimatic implications [Abstr.]. Geol. Soc. Amer. Abstr. with Programs, 3(2):190.

STAUFFER, P. H.
1967. Sedimentologic evidence on Eocene correlations, Santa Ynez Mountains, California. Amer. Assoc. Petrol. Geol. Bull., 51(4):607–611, 3 text figs.

STEINECK, P. L., and J. M. GIBSON
1971. Age and correlation of Eocene Ulatisian and Narizian stages, California. Geol. Soc. Amer. Bull., 82:477–480.

STENZEL, H. B.
1971. Mollusca 6 (part N), vol. 3. *In* R. C. Moore, ed., Treatise on invertebrate paleontology. Lawrence: University of Kansas Press. Pp. 953–1224.

STEWART, R. B.
1926. Gabb's California fossil type gastropods. Acad. Nat. Sci. Phila. Proc., 78:287–447, pls. 20–32.
1930. Gabb's California Cretaceous and Tertiary type lamellibranchs. Acad. Nat. Sci. Phila. Spec. Publ. 3. 314 pp., 17 pls.
1946. Geology of Reef Ridge, Coalinga District, California. U.S. Geol. Surv. Prof. Paper 205-C. Pp. 81–115, pls. 9–17, figs. 10–13, 2 tables.
1949. Lower Tertiary stratigraphy of Mount Diablo, Marysville Buttes, and west border of lower Central Valley of California. U.S. Geol. Surv. Oil and Gas Investigations Preliminary Chart 34. Sheet 2.

TALIAFERRO, K. L.
1924. Notes on the geology of Ventura County, California. Amer. Assoc. Petrol. Geol. Bull., 8:789–810.

TEGLAND, N. M.
1933. The fauna of the type Blakeley, upper Oligocene of Washington. Univ. Calif. Publ. Bull. Dept. Geol. Sci., 23(3):81–174, pls. 2–15, 2 maps.

TURNER, F. E.
1938. Stratigraphy and Mollusca of the Eocene of western Oregon. Geol. Soc. Amer. Spec. Paper 10. 117 pp., 22 pls.

VEDDER, J. G.
1972. Revision of stratigraphic names for some Eocene Formations in Santa Barbara and Ventura counties, California. U.S. Geol. Surv. Bull. 1354-D. Pp. 1–12.

VEDDER, J. G., and R. D. BROWN, JR.
1968. Structural and stratigraphic relations along the Nacimiento Fault in the southern Santa Lucia Range and San Rafael Mountains, California. *In* W. R. Dickinson and A. Grantz, eds., Proceedings of Conference on Geologic Problems of San Andreas Fault System. Stanford Univ. Publ. Geol. Sci., 11:242–259.

VERASTEGUI, P.
1953. The pelecypod genus *Venericardia* in the Paleocene and Eocene of western North America. Palaeontographica Americana, vol. 3, no. 25. 112 p., 22 pls.

VOKES, E. H.
1968. Cenozoic Muricidae of the Western Atlantic region. Part 4. *Hexaplex* and *Murexiella*. Tulane Studies in Geol. 6 (3):85–126, pls. 1–8.
1971. Catalogue of the genus Murex Linné (Mollusca: Gastropoda); Muricinae, Ocenebrinae. Bull. Amer. Paleontol., vol. 61, no. 268. 141 pp.

VOKES, H. E.
1935*a*. Notes on the variation and synonymy of *Ostrea idriaensis* Gabb. Univ. Calif. Publ. Bull. Dept. Geol. Sci., 23(9):291–304, pls. 22–24.
1935*b*. The genus Velates in the Eocene of California. Univ. Calif. Publ. Bull. Dept. Geol. Sci., 23(12):381–390, pls. 25, 26.
1939. Molluscan faunas of the Domengine and Arroyo Hondo Formations of the California Eocene. New York Acad. Sci. Annals, vol. 38. 246 pp., 22 pls.

WARING, C. A.
1917. Stratigraphic and faunal relations of Martinez to the Chico and Tejon of southern California. Calif. Acad. Sci. Proc., 4th ser., 7(4):41–124, pls. 7–16.

WARME, J. E.

1971. Paleoecological aspects of a modern coastal lagoon. Univ. Calif. Publ. Geol. Sci., vol. 87. 131 pp., 9 pls., 4 figs., 3 maps, 1 table.

WARMKE, G. L., and R. T. ABBOTT

1962. Caribbean seashells: A guide to the marine mollusks of Puerto Rico and other West Indian Islands, Bermuda, and the lower Florida Keys. Narberth, Pa.: Livingston Publishing Co. 348 pp., 44 pls.

WATTS, W. L.

1897. Oil and gas yielding formations of the Los Angeles, Ventura and Santa Barbara Counties, California. Calif. State Mining Bur. Bull. 11. Pp. 1–94.

1901. Oil and gas yielding formations of California. Calif. Mining Bur. Bull. 19. 236 pp.

WEAVER, C. E.

1912. A preliminary report on the Tertiary paleontology of western Washington. Wash. Geol. Surv. Bull., 15:1–80, pls. 1–15.

1916. Tertiary faunal horizons of western Washington. Univ. Wash. Publ. Geol., 1(1):1–67, pls. 1–5.

1942. Paleontology of the marine Tertiary formations of Oregon and Washington. Univ. Wash. Publ. Geol., 5:1–274, pls. 1–104.

1949. Geology of the Coast Ranges immediately north of the San Francisco Bay region, California. Geol. Soc. Amer. 35. 242 pp.

1953. Eocene and Paleocene deposits at Martinez, California. Univ. Wash. Publ. Geol., vol. 7. 102 pp., illus.

WEAVER, C. E., and K. V. PALMER

1922. Fauna from the Eocene of Washington. Univ. Wash. Publ. Geol., 1(3):1–56, pls. 8–12.

WEAVER, C. E., et al.

1944. Correlation of the marine Cenozoic formations of western North America. Geol. Soc. Amer. Bull., 55:569–598, 1 pl.

WEIMER, R. J., and J. H. HOYT

1964. Burrows of *Callianassa major* Say, geologic indicators of littoral and shallow neritic environments. J. Paleontol. 38(4):761–767, pls. 123, 124.

WELDAY, E. E.

1960. Geology of the San Guillermo Mountain area, California. Unpub. M.A. thesis, Pomona College, Pomona, Calif.

WENZ, WILHELM

1938–1944. Handbuch der Paläeozoologie. Berlin: Gebrüder Borntraeger. 8 vols.

WHITE, C. A.

1889. On invertebrate fossils from the Pacific Coast. U.S. Geol. Surv. Bull. 51. 102 pp., 14 pls.

WHITE, R. T.

1938. The Eocene Lodo Formation and Cerros member of California [Abstr.]. Geol. Soc. Amer. Proc. Pp. 256–257.

WOODRING, W. P.

1931. Age of the orbitoid-bearing limestone and *Turritella variata* zone of the western Santa Ynez Range, California. San Diego Soc. Nat. History Trans., 6(25):371–388.

PLATES

PLATE 1

Fig. 1. *Acila* (*Trunacila*) *decisa* (Conrad). × 2. Specimen UCR 4675/231, locality UCR 4675.

Fig. 2. *Ledina fresnoensis* (Dickerson). × 2. Specimen UCR 4662/101, locality UCR 4662.

Fig. 3. *Nuculana* (*Saccella*) gabbii (Gabb). × 2. Specimen UCR 4743/301, locality UCR 4743.

Fig. 4. *Nuculana* (*Saccella*) *hondana* Vokes. × 2. Specimen UCR 4662/12, locality UCR 4662.

Fig. 5. *Glycymeris* (*Glycymeris*) n. sp.? aff. *G.* (*G.*) *perrini* Dickerson. × 2. Specimen UCR 4697/301, locality UCR 4697.

Fig. 6. *Glycymeris* (*Glycymeris*) *viticola* Anderson and Hanna. × 2. Specimen UCR 4723/141, locality UCR 4723.

Fig. 7. *Limopsis* (*Limopsis*) *marysvillensis* (Dickerson). × 2. Specimen UCR 4656/101, locality UCR 4656.

Fig. 8. *Arca* (*Arca*) n. sp.? × 2. Specimen UCR 4667/131, locality UCR 4667.

Fig. 9. *Nayadina* (*Exputens*) *llajasensis* (Clark). × 1. Specimen UCR 4659/81, locality UCR 4659.

Fig. 10. *Pteria pellucida* (Gabb). × 1. Specimen UCR 4714/6, locality UCR 4714.

Figs. 11–13. *Odontogryphaea? haleyi* (Hertlein). × 1. (11) specimen UCR 4670/32, locality UCR 4670; (12) specimen UCR 4670/33, locality UCR 4670; (13) specimen UCR 4670/34, locality UCR 4670.

Fig. 14. *Glycymeris* (*Glycymeris*) aff. *G.* (*G.*) *fresnoensis* Dickerson. × 2. Specimen UCR 4661/91, locality UCR 4661.

Fig. 15. *Claibornites diegoensis* (Dickerson). × 1. Specimen UCR 4752/41, locality UCR 4752.

Fig. 16. *Venericardia* (*Pacificor*) *hornii lutmani* Turner. Umbonal view, × 1. Specimen UCR 4662/40, locality UCR 4662.

Fig. 17. *Acanthocardia* (*Schedocardia*) *brewerii* (Gabb). × 1. Specimen UCR 4741/4, locality UCR 4741.

1
2
3
4
5
6
7
8
9
10
11
12
13
14
15
16
17

PLATE 2

Fig. 1. *Solena* (*Eosolen*) *coosensis* Turner. × 1. Specimen UCR 4676/22, locality UCR 4676.

Fig. 2. *Tellina castacana* Anderson and Hanna. × 2. Specimen UCR 4743/502, locality UCR 4743.

Fig. 3. *Tellina* n. sp.? aff. *T. townsendensis* Clark. × 1. Specimen UCR 4743/6, locality UCR 4743.

Fig. 4. *Macoma viticola* Anderson and Hanna. × 2. Specimen UCR 4726/12, locality UCR 4726.

Fig. 5. *Spisula bisculpturata* Anderson and Hanna. × 1. Specimen UCR 4743/701, locality UCR 4743.

Fig. 6. *Isognomon* n. sp.? × 0.5. Specimen UCR 4752/61, locality UCR 4752.

Fig. 7. *Tellina soledadensis* Hanna. × 1. Specimen UCR 4678/16, locality UCR 4678.

Fig. 8. *Venericardia* (*Pacificor*) *hornii lutmani* Turner. Left valve, × 1. Specimen UCR 4662/40, locality UCR 4662.

1
2
3
4
5
6
7
8

PLATE 3

Figs. 1, 4. *Pitar* (*Lamelliconcha*) *avenalensis* Vokes. × 2. (1) specimen UCR 4679/31, locality UCR 4679; (4) specimen UCR 4679/30, locality UCR 4679.

Fig. 2. *Pitar* (*Lamelliconcha*) *soledadensis* (Hanna). × 1. Specimen UCR 4705/1, locality UCR 4705.

Fig. 3. *Venericardia* (*Pacificor*) *hornii* (Gabb) s.s. × 1. Specimen UCR 4745/1, locality UCR 4745.

Fig. 5. *Callista* (*Macrocallista*) *andersoni* (Dickerson). × 1. Specimen UCR 4731/101, locality UCR 4731.

Fig. 6. *Callista* (*Costacallista*) *hornii* (Gabb). × 1. Specimen UCR 4723/ 61, locality UCR 4723.

Fig. 7. *Pitar* (*Lamelliconcha*) *joaquinensis* Vokes. × 2. Specimen UCR 4673/201, locality UCR 4673.

Figs. 8, 11. *Pelecyora gabbi* (Arnold). × 1. (8) specimen UCR 4683/701, locality UCR 4683; (11) specimen UCR 4683/702, locality UCR 4683.

Fig. 9. *Cyclinella elevata* (Gabb). × 1. Specimen UCR 4726/21, locality UCR 4726.

Fig. 10. *Pitar* (*Lamelliconcha*) *dickersoni* Givens, new species. Specimen UCR 4721/171, locality UCR 4721.

1
2
3
4
5
6
7
8
9
10
11

PLATE 4

Fig. 1. *Venericardia* (*Pacificor*) *hornii* (Gabb) cf. subsp. *calafia* Stewart. × 1. Specimen UCR 4684/6, locality UCR 4684.

Fig. 2. *Pitar* (*Pitar*) cf. *P.* (*P.*) *lascrucensis* Kleinpell and Weaver. × 1. Specimen UCR 4732/101, locality UCR 4732.

Fig. 3. *Cardiomya israelskyi* (Hanna). × 2. Specimen UCR 4700/2, locality UCR 4700.

Fig. 4. *Thracia* (*Thracia*) *dilleri* Dall. × 1. Specimen UCR 4731/81, locality UCR 4731.

Fig. 5. *Pholadomya* n. sp. × 2. Specimen UCR 4662/110, locality UCR 4662.

Fig. 6. *Corbula* (*Caryocorbula*) *hornii* Gabb. × 2. Specimen UCR 4717/301, locality UCR 4717.

Fig. 7. *Corbula* (*Caryocorbula*) *dickersoni* Weaver and Palmer. × 2. Specimen UCR 4707/32, locality UCR 4707.

Fig. 8. *Periploma* n. sp.? × 1.3. Specimen UCR 4732/2, locality UCR 4732.

Fig. 9. *Corbula* (*Caryocorbula*) *parilis* Gabb. × 3. Specimen UCR 4680/101, locality UCR 4680.

1
2
3
4
5
6
7
8
9

PLATE 5

Figs. 1, 2. *Solariella crenulata* (Gabb). Specimen UCR 4681/201, locality UCR 4681. (1) Apertural view, × 2; (2) Dorsal view, × 2.5.

Fig. 3. *Homalopoma umpquaensis* Merriam and Turner. × 2. Specimen UCR 4676/6, locality UCR 4676.

Fig. 4. *Nerita* (*Theliostyla*) *triangulata* Gabb. × 2. Specimen UCR 4747/3, locality UCR 4747.

Figs. 5, 6, 13. *Velates perversus* (Gmelin). (5) Dorsal view, × 1. Specimen UCR 4668/33, locality UCR 4668; (6) Ventral view, × 1. Specimen UCR 4668/34, locality UCR 4668; (13) Side view, × 1. Specimen UCR 4668/33, locality UCR 4668.

Figs. 7–10. *Turritella andersoni* Dickerson s.s. (7) specimen UCR 4659/203, locality UCR 4659. × 2; (8) specimen UCR 4659/201, locality UCR 4659. × 2; (9) specimen UCR 4659/202, locality UCR 4659. × 2; (10) specimen UCR 4667/201, locality UCR 4667. × 1.

Figs. 11, 12, 14. *Turritella andersoni lawsoni* Dickerson. (11) specimen UCR 4701/11, locality UCR 4701. × 1; (12) specimen UCR 4700/10, locality UCR 4700. × 1; (14) specimen UCR 4694/10, locality UCR 4694. × 2. Incomplete specimen showing typical adult whorl profile.

Fig. 15. *Turritella buwaldana* Dickerson s.s. × 2. Specimen UCR 4705/201, locality UCR 4705.

Figs. 16, 17. *Turritella buwaldana crooki* Merriam and Turner. (16) specimen UCR 4656/201, locality UCR 4656. × 2; (17) specimen UCR 4656/202, locality UCR 4656. × 2.

1
2
3
4
5
6
7
8
9
10
11
12
13
14
15
16
17

PLATE 6

Fig. 1. *Turritella scrippsensis* Hanna. × 2. Specimen UCR 4690/10, locality UCR 4690.

Fig. 2. *Turritella uvasana* Conrad s.s. × 1. Specimen UCR 4722/210, locality UCR 4722.

Figs. 3, 4. *Turritella uvasana applinae* Hanna. (3) specimen UCR 4694/21, locality UCR 4694. × 1; (4) specimen UCR 4702/101, locality UCR 4702. × 1.

Figs. 5–7. *Turritella uvasana infera* Merriam. (5) specimen UCR 4662/201, locality UCR 4662. × 1; (6) specimen UCR 4661/41, locality UCR 4661. × 1; (7) specimen UCR 4659/103, locality UCR 4659. × 1.

Figs. 8–11. *Turritella uvasana neopleura* Merriam. (8) specimen UCR 4706/201, locality UCR 4706. × 1; (9) specimen UCR 4706/204, locality UCR 4706. × 1; (10) specimen UCR 4744/21, locality UCR 4744. × 1; (11) specimen UCR 4743/2, locality UCR 4743. × 1.

Figs. 12–16. *Turritella uvasana sargeanti* Anderson and Hanna. (12) specimen UCR 4726/101, locality UCR 4726. × 1; (13) specimen UCR 4726/103, locality UCR 4726. × 1; (14) specimen UCR 4726/102, locality UCR 4726. × 1; (15) specimen UCR 4732/301, locality UCR 4732. × 1; (16) specimen UCR 4732/302, locality UCR 4732. × 1.

Fig. 17. *Loxotrema turritum* Gabb. × 1. Specimen UCR 4747/401, locality UCR 4747.

Fig. 18. "*Trichotropis*" *lajollaensis* Hanna. × 1. Specimen UCR 4747/501, locality UCR 4747.

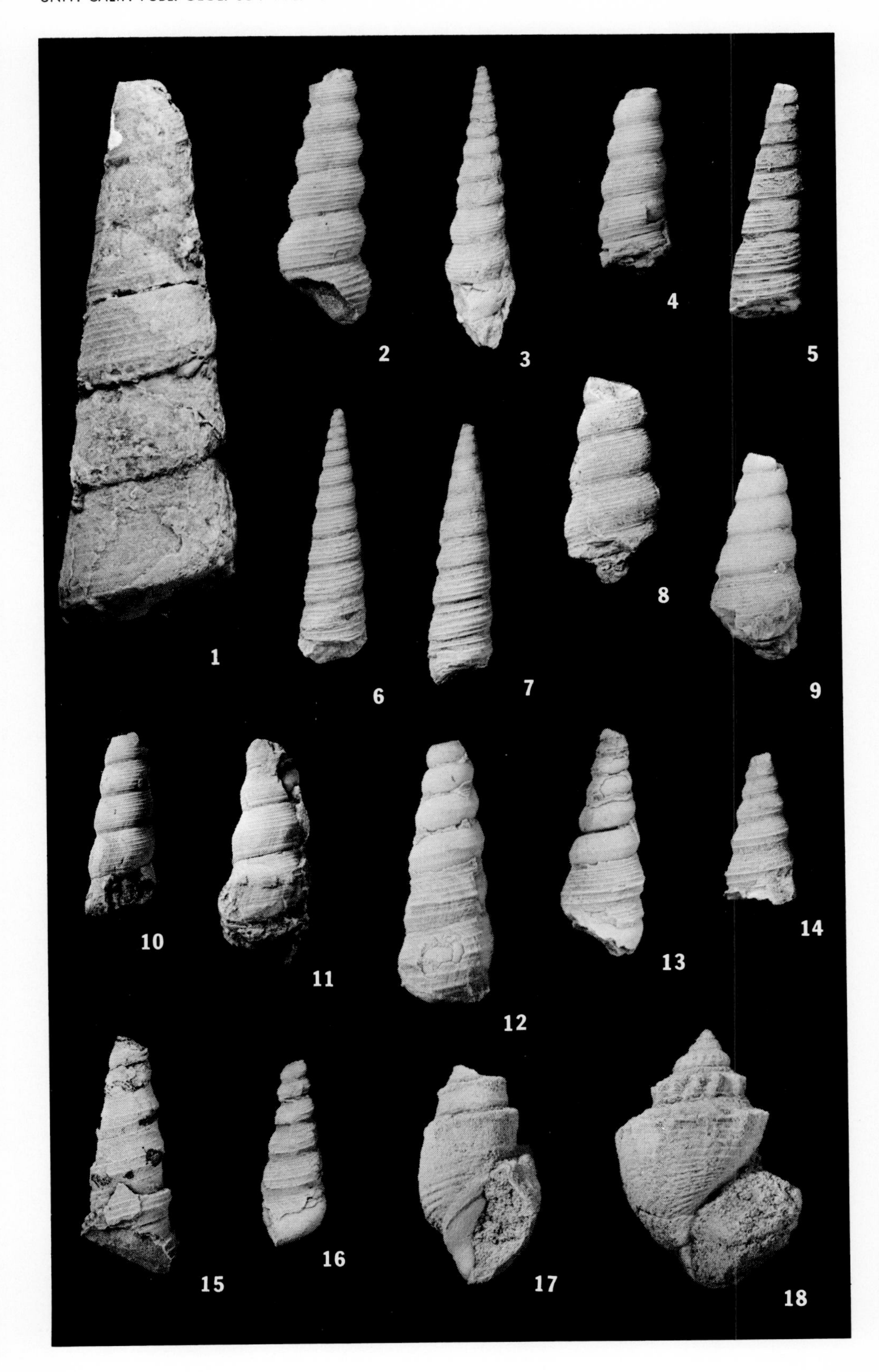
1
2
3
4
5
6
7
8
9
10
11
12
13
14
15
16
17
18

PLATE 7

Figs. 1–3. *Architectonica (Stellaxis) cognata* Gabb. × 2. Specimen UCR 4750/20, locality UCR 4750.

Fig. 4. *Cerithium cliffensis* Hanna. × 2. Specimen UCR 4703/30, locality UCR 4703.

Figs. 5–7. *Architectonica (Architectonica) hornii* Gabb. × 2. Specimen UCR 4741/60, locality UCR 4741.

Fig. 8. *Xenophora stocki* Dickerson. × 1. Specimen UCR 4680/7, locality UCR 4680.

Fig. 9. *Architectonica (Solariaxis) ullreyana* Dickerson. × 2. Specimen UCR 4690/9, locality UCR 4690. Dorsal view, showing crenulate spiral ribbing.

Fig. 10. *Campanilopa dilloni* Hanna and Hertlein. × 1. Specimen UCR 4668/101, locality UCR 4668.

Fig. 11. *Ectinochilus (Cowlitzia) canalifer* (Gabb). × 2. Specimen UCR 4714/30, locality UCR 4714.

Fig. 12. *Ectinochilus (Cowlitzia) supraplicatus* (Gabb). × 2. Specimen UCR 4753/131, locality UCR 4753.

Figs. 13, 16. *Ectinochilus (Macilentos) macilentus* (White). (13) specimen UCR 4680/17, locality UCR 4680. × 2; (16) specimen UCR 4679/7, locality UCR 4679. × 2.

Fig. 14. *Euspira nuciformis* (Gabb). × 1. Specimen UCR 4671/2, locality UCR 4671.

Figs. 15, 17. *Euspira clementensis* (Hanna). × 2. Specimen UCR 4688/25, locality UCR 4688. (15) dorsal view; note channeled suture; (17) apertural view.

Fig. 18. *Natica (Natica) uvasana* Gabb. × 2. Specimen UCR 4715/4, locality UCR 4715.

Fig. 19. *Turritella uvasana applinae* Hanna. × 2. Specimen UCR 4688/101, locality UCR 4688.

1
2
3
4
5
6
7
8
9
10
11
12
13
14
15
16
17
18
19

PLATE 8

Fig. 1. *Natica (Natica) uvasana* Gabb. × 2. Specimen UCR 4714/12, locality UCR 4714.

Fig. 2. *Polinices hornii* (Gabb). × 1. Specimen UCR 4720/2, locality UCR 4720.

Fig. 3. *Crommium andersoni* (Dickerson). × 1. Specimen UCR 4659/5, locality UCR 4659.

Fig. 4. *Galeodea (Gomphopages) sutterensis* Dickerson. × 1. Specimen UCR 4658/5, locality UCR 4658.

Fig. 5. *Tejonia lajollaensis* (Stewart). × 2. Specimen UCR 4703/151, locality UCR 4703. Note absence of carina on shoulder of whorl.

Figs. 6, 10. *Pachycrommium? clarki* (Stewart). (6) specimen UCR 4699/101, locality UCR 4699. × 1; (10) specimen UCR 4674/10, locality UCR 4674. × 1.

Fig. 7. *Coalingodea tuberculiformis* (Hanna). × 1. Specimen UCR 4752/1, locality UCR 4752. Note beaded spiral ribs.

Fig. 8. *Tejonia moragai* (Stewart). × 2. Specimen UCR 4706/101, locality UCR 4706. Note carinate shoulder.

Figs. 9, 11. *Pachycrommium?* n. sp. × 1. Specimen UCR 4668/11, locality UCR 4668.

Fig. 12. *Natica (Carinacca) rosensis* Hanna. × 2. Specimen UCR 4703/7, locality UCR 4703. Apertural view, showing basal funicle.

Fig. 13. *Ampullella hewitti* Hanna and Hertlein. × 1. Specimen UCR 4668/21, locality UCR 4668.

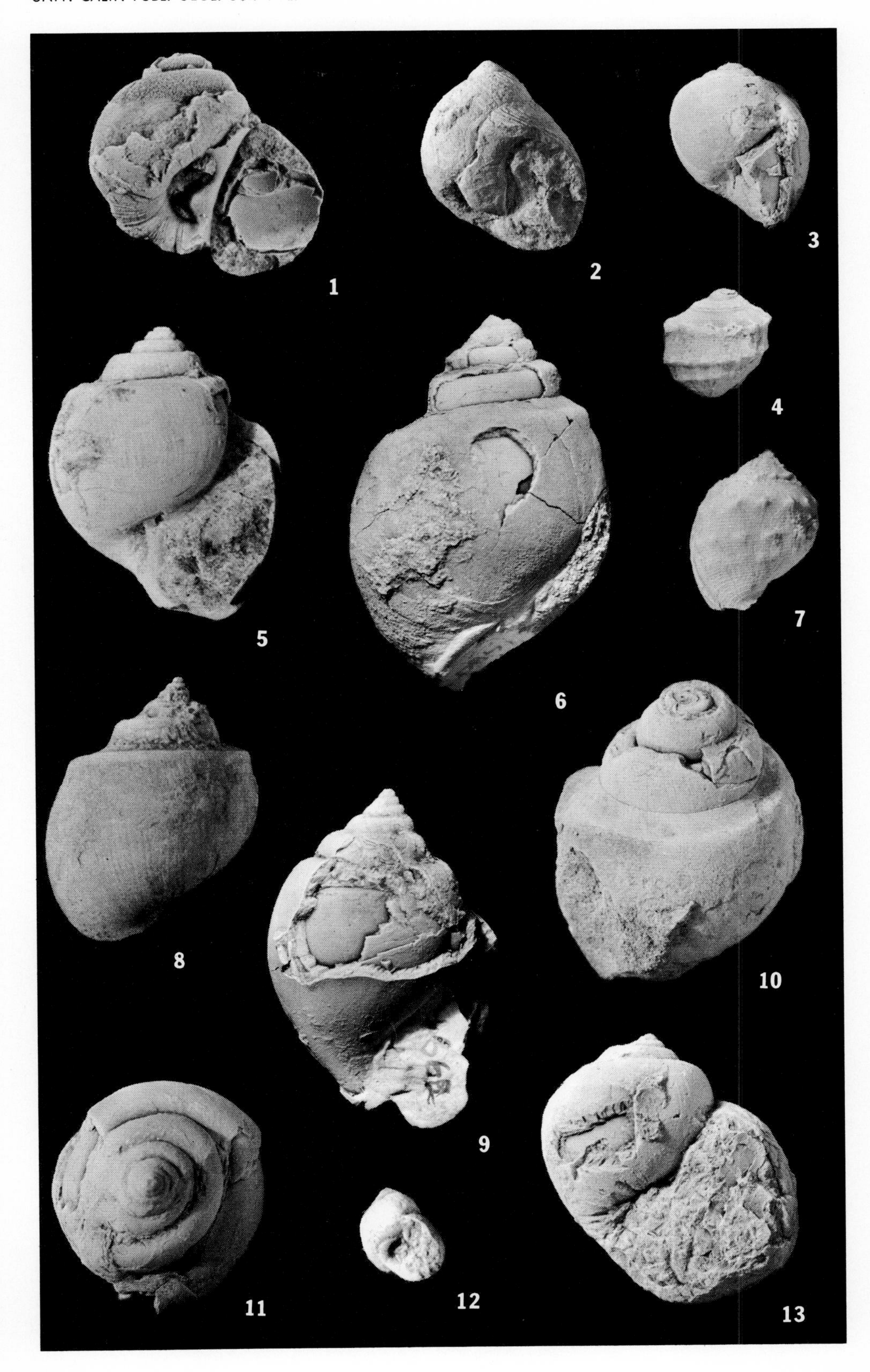
1
2
3
4
5
6
7
8
9
10
11
12
13

PLATE 9

Figs. 1, 3. *Globularia (Eocernina) hannibali* (Dickerson). (1) specimen UCR 4699/18, locality UCR 4699. × 1; (3) specimen UCR 4658/11, locality UCR 4658. × 1.

Fig. 2. *Ficus mamillata* Gabb. × 1. Specimen UCR 4715/9, locality UCR 4715.

Figs. 4, 5. *Olequahia domenginica* (Vokes). (4) specimen UCR 4686/8, locality UCR 4686. × 1; (5) specimen UCR 4679/1, locality UCR 4679. × 1.

Fig. 6. *Pseudoliva inornata* Dickerson × 2. Specimen UCR 4707/31, locality UCR 4707.

Figs. 7, 9. *Ficopsis cooperiana* Stewart. (7) specimen UCR 4679/111, locality UCR 4679. × 2; (9) specimen UCR 4703/14, locality UCR 4703. × 1.

Fig. 8. *Siphonalia sopenahensis* (Weaver). × 2. Specimen UCR 4719/16, locality UCR 4719.

Fig. 10. *Ficopsis hornii* (Gabb). × 1. Specimen UCR 4741/201, locality UCR 4741.

Fig. 11. *Ficopsis remondii crescentensis* Weaver and Palmer. × 1. Specimen UCR 4703/13, locality UCR 4703.

Fig. 12. *Ranellina pilsbryi* Stewart. × 2. Specimen UCR 4721/112, locality UCR 4721.

Fig. 13. *Hexaplex? whitneyi* (Gabb). × 5½. Specimen UCR 4715/3, locality UCR 4715.

Fig. 14. *Pseudoliva volutaeformis* Gabb. × 2. Specimen UCR 4706/151, locality UCR 4706.

1
2
3
4
5
6
7
8
9
10
11
12
13
14

PLATE 10

Fig. 1. *Molopophorus antiquatus* (Gabb). × 2. Specimen UCR 4707/301, locality UCR 4707.

Fig. 2. *Molopophorus tejonensis* Dickerson. × 2. Specimen UCR 4722/401, locality UCR 4722.

Fig. 3. *Brachysphingus mammilatus* Clark and Woodford. × 1. Specimen UCR 4662/5, locality UCR 4662.

Figs. 4, 5. *Clavilithes tabulatus* (Dickerson). (4) specimen UCR 4661/5, locality UCR 4661. × 1; (5) specimen UCR 4658/7, locality UCR 4658. × 1.

Figs. 6, 7. *Fusinus teglandae* Hanna. (6) specimen UCR 4684/5, locality UCR 4684. × 1; (7) specimen UCR 4700/4, locality UCR 4700. × 1.

Fig. 8. *Falsifusus* cf. *F. marysvillensis* (Merriam and Turner). × 2. Specimen UCR 4659/9, locality UCR 4659.

Fig. 9. *Perse martinez* (Gabb). × 2. Specimen UCR 4707/201, locality UCR 4707.

Fig. 10. *Strepsidura ficus* (Gabb). × 2. Specimen UCR 4721/131, locality UCR 4721.

Fig. 11. *Acteon quercus* Anderson and Hanna. × 5. Specimen UCR 4722/14, locality UCR 4722.

Fig. 12. *Conus hornii* Gabb. × 2. Specimen UCR 4731/14, locality UCR 4731.

Fig. 13. *Perse sinuata* (Gabb). × 2. Specimen UCR 4719/2, locality UCR 4719.

Fig. 14. *Conomitra* aff. *C. washingtoniana* (Weaver). × 2. Specimen UCR 4706/26, locality UCR 4706.

Figs. 15, 16. *Pseudoperissolax blakei praeblakei* Vokes. (15) specimen UCR 4682/7, locality UCR 4682. × 1; (16) specimen UCR 4685/121, locality UCR 4685. × 1.

Fig. 17. *Pseudoperissolax blakei* (Conrad) s.s. × 1. Specimen UCR 1807/15, locality UCR 1807, type Tejon Formation, San Joaquin Valley, California.

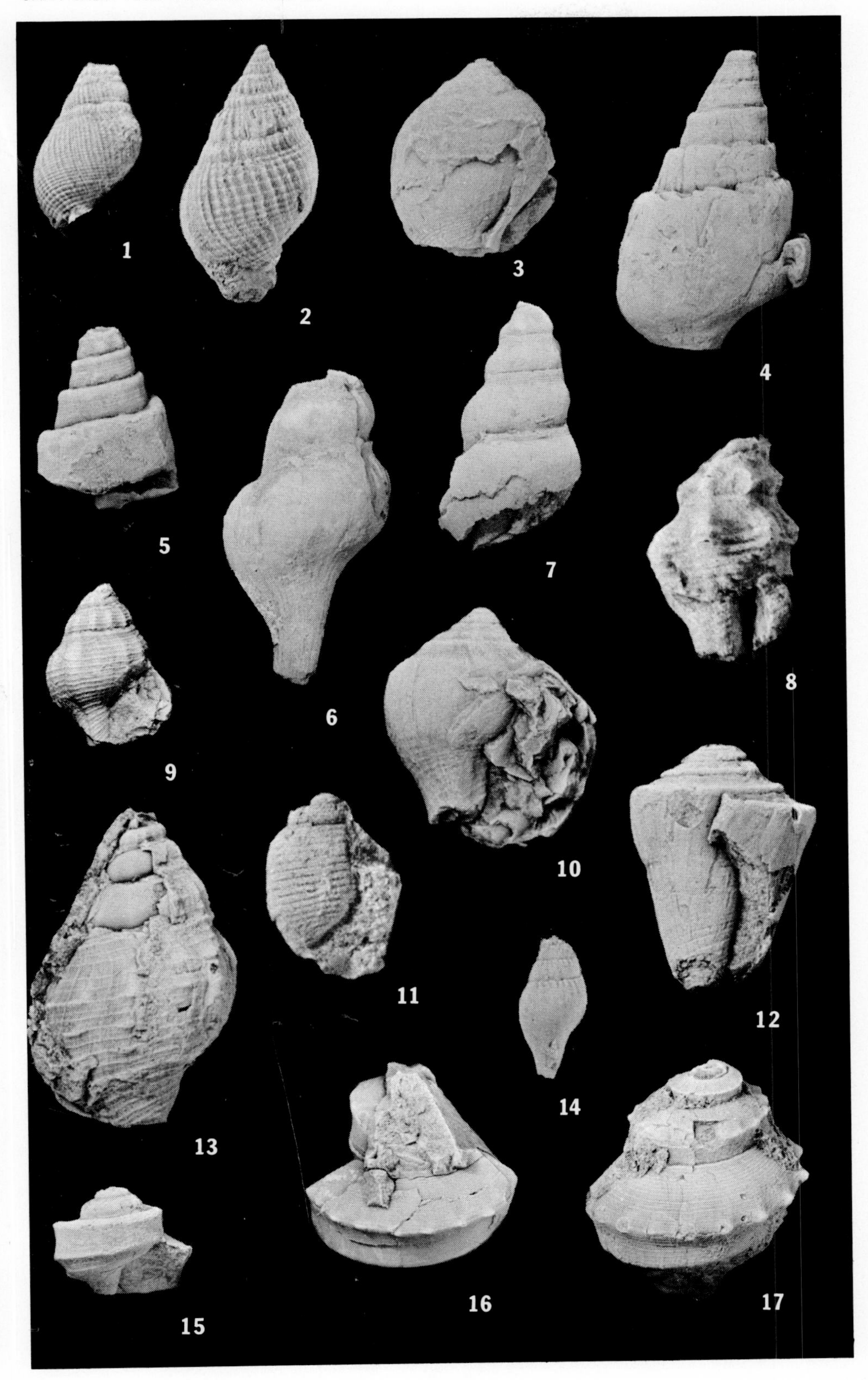
1
2
3
4
5
6
7
8
9
10
11
12
13
14
15
16
17

PLATE 11

Fig. 1. *Conus* n. sp.? aff. *C. californianus* (Conrad). × 5. Specimen UCR 4706/141, locality UCR 4706.

Fig. 2. *Exilia fausta* Anderson and Hanna. × 5. Specimen UCR 4726/19, locality UCR 4726.

Fig. 3. *Nekewis io* (Gabb). × 2. Specimen UCR 4741/11, locality UCR 4741.

Figs. 4, 11. *Turricula praeattenuata* (Gabb). (4) specimen UCR 4676/4, locality UCR 4676. × 2; (11) specimen UCR 4686/6, locality UCR 4686. × 5.

Figs. 5, 7. *Surculites mathewsonii* (Gabb). (5) specimen UCR 4679/4, locality UCR 4679. × 1; (7) specimen UCR 4682/2, locality UCR 4682. × 1.

Fig. 6. *Turricula cohni* (Dickerson). × 2. Specimen UCR 4741/251, locality UCR 4741.

Figs. 8, 10. *Sassia bilineata* (Dickerson). × 2. Specimen UCR 4707/23, locality UCR 4707.

Fig. 9. *Pleurofusia fresnoensis* (Arnold). × 5. Specimen UCR 4706/231, locality UCR 4706.

1
2
3
4
5
6
7
8
9
10
11